Management of Noise Pollution

Edited by Mia Suhanek

Published in London, United Kingdom

Management of Noise Pollution
http://dx.doi.org/10.5772/intechopen.106098
Edited by Mia Suhanek

Contributors
Alice Elizabeth González, Martha Cobo Dorado, Gissell Rodríguez Milan, Mariana Alves-Pereira, Huub Bakker, Richard Mann, Philip Dickinson, Rachel Summers, Aleksey Bogomolov, Sergey Dragan, Dietrich Schwela, Mia Suhanek

First published in London, United Kingdom, 2023 by IntechOpen
IntechOpen is the global imprint of INTECHOPEN LIMITED, registered in England and Wales, registration number: 11086078, 5 Princes Gate Court, London, SW7 2QJ, United Kingdom
Printed in Croatia

British Library Cataloguing-in-Publication Data
A catalogue record for this book is available from the British Library

Additional hard and PDF copies can be obtained from orders@intechopen.com

Management of Noise Pollution
Edited by Mia Suhanek
p. cm.

This title is part of the Environmental Sciences Book Series, Volume 3
Topic: Pollution
Series Editor: J. Kevin Summers
Topic Editors: Ismail Md. Mofizur Rahman and Zinnat Ara Begum

Print ISBN 978-1-83768-655-1
Online ISBN 978-1-83768-656-8
eBook (PDF) ISBN 978-1-83768-657-5
ISSN 2754-6713

IntechOpen Book Series

Environmental Sciences

Volume 3

Aims and Scope of the Series

Scientists have long researched to understand the environment and man's place in it. The search for this knowledge grows in importance as rapid increases in population and economic development intensify humans' stresses on ecosystems. Fortunately, rapid increases in multiple scientific areas are advancing our understanding of environmental sciences. Breakthroughs in computing, molecular biology, ecology, and sustainability science are enhancing our ability to utilize environmental sciences to address real-world problems.

The four topics of this book series - Pollution; Environmental Resilience and Management; Ecosystems and Biodiversity; and Water Science - will address important areas of advancement in the environmental sciences. They will represent an excellent initial grouping of published works on these critical topics.

Meet the Series Editor

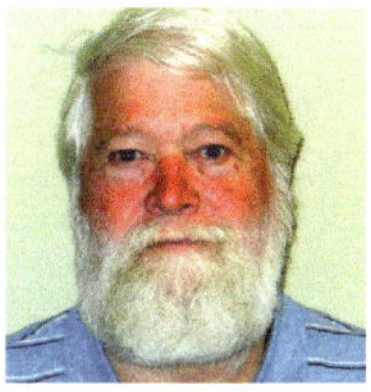

J. Kevin Summers is a Senior Research Ecologist at the Environmental Protection Agency's (EPA) Gulf Ecosystem Measurement and Modeling Division. He is currently working with colleagues in the Sustainable and Healthy Communities Program to develop an index of community resilience to natural hazards, an index of human well-being that can be linked to changes in the ecosystem, social and economic services, and a community sustainability tool for communities with populations under 40,000. He leads research efforts for indicator and indices development. Dr. Summers is a systems ecologist and began his career at the EPA in 1989 and has worked in various programs and capacities. This includes leading the National Coastal Assessment in collaboration with the Office of Water which culminated in the award-winning National Coastal Condition Report series (four volumes between 2001 and 2012), and which integrates water quality, sediment quality, habitat, and biological data to assess the ecosystem condition of the United States estuaries. He was acting National Program Director for Ecology for the EPA between 2004 and 2006. He has authored approximately 150 peer-reviewed journal articles, book chapters, and reports and has received many awards for technical accomplishments from the EPA and from outside of the agency. Dr. Summers holds a BA in Zoology and Psychology, an MA in Ecology, and Ph.D. in Systems Ecology/Biology.

Meet the Volume Editor

Mia Suhanek received her doctoral degree in 2013 with a dissertation entitled "Evaluation of soundscapes regarding sudden and unexpected sound changes." Over the years, she has participated in numerous scientific congresses and has been involved in scientific research, teaching, and professional work in electroacoustics, noise, audiotechnics, and digital logic. Currently, Dr. Suhanek is an assistant professor in the Department of Electroacoustics, Faculty of Electrical Engineering and Computing, University of Zagreb, Croatia. Her research interests include soundscape modeling, analysis of noise pollution, and psychoacoustics.

Contents

Preface

This book raises awareness about noise pollution and its negative impact on the overall quality of life, especially in urban areas. Noise pollution, when compared to other environmental pollutions, is often neglected. A possible reason for this lies in its accumulating character, that is, the negative consequences of noise exposure only appear after long-term exposure.

Each chapter in this book draws attention to a specific concern, analyzing it and offering possible solutions.

Chapter 1 serves as an introduction to noise pollution, its negative impact, and two possible approaches to dealing with it: noise barriers and the soundscape concept. It critically analyzes each of these options and presents both pros and cons.

Chapter 2 deals with the challenges of environmental noise policies and governance, noise emissions, noise transmission modeling, and health and economic risk assessment in developing countries (e.g., Asia, Africa, and Latin America). Although appropriate legislation and laws are developed and provided, their enforcement is rarely implemented. Therefore, the chapter provides guidelines for a strategic framework to overcome these challenges and enable countries to attain sustainable environmental noise management.

Chapter 3 summarizes the methodological aspects of monitoring industrial and transport noise, including the main physical characteristics, features of sources, measuring instruments, features of hygienic regulation of industrial and transport noise, and means and methods of protection against it.

Chapter 4 presents a case study evaluating the acoustic performance of a tree barrier. Green noise barriers are becoming a propulsive acoustic instrument for noise reduction. Today, they must satisfy both aesthetic and noise reduction requirements. In addition to noise reduction, it has been proven that green barriers increase the quality of air in urban areas.

Chapter 5 studies infrasonic ($\leq$ 20 Hz) noise exposure in a residential area in the vicinity of wind power plants. Infrasound by its definition should be inaudible to humans, however, studies have shown that the highest peaks of the wind turbine acoustic signature (up to 25 dB over background noise at 0.5–5 Hz) appears to trigger severe biological reactions. Therefore, the chapter suggests a new methodology.

We hope that this book will give readers new insight into the problems of noise pollution today as well as provide new ideas for possible methods to reduce noise levels and therefore improve the overall quality of life, especially for those residing in cities and urban areas.

Mia Suhanek
Faculty of Electrical Engineering and Computing,
University of Zagreb,
Zagreb, Croatia

Chapter 1

Introductory Chapter: Management of Noise Pollution

Mia Suhanek

1. Introduction

Noise is defined as any disturbing or unwanted sound that influences or deteriorates human or wildlife [1]. Although noise constantly surrounds us, noise pollution generally receives less attention than other environmental pollutants (e.g., water pollution, soil pollution, air pollution, etc.) [2]. This can be explained with the fact that noise exposure has an accumulating character which means that the negative impact of noise can be detected after a long period of time. Long exposure to noise pollution can cause bad mood, fatigue, insomnia, headache, loss of concentration, reduced work ability and finally the worst possible case-scenario permanent hearing impairment [3–5]. In addition, recent research studies unfortunately show that environmental noise has an impact on several cardiovascular (e.g., increased blood pressure) and metabolic effects, cognitive impairment among children, annoyance, stress-related mental health risks and tinnitus [6–8].

When discussing noise in general, one also needs to keep in mind that a certain sound perceived as desired or wanted by one person can be perceived as noise for someone else. This can be a devious task when analysing noise and implementing solutions for noise reduction.

Human ear can hear a relatively large ratio of the effective maximum and minimum values of the sound pressure which are expressed then in decibels (dB). Sound level is expressed in decibels in relation to the reference sound pressure level (Pa) which corresponds to the threshold of audibility of the average person at 1 kHz (**Figure 1**) [9].

In addition, noise can be described with noise perception parameters such as loudness (son), sharpness (acum), roughness (asper), fluctuation strength (vacil) and psychoacoustic annoyance (son) [10].

When dealing with the management of noise pollution, i.e., reduction of noise pollution, it has been proven that an interdisciplinary approach is required. From acoustical point of view, a traditional approach to reduction of noise pollution is noise barriers, while a more modern and propulsive approach is the soundscape concept.

Noise barrier is a sound "obstacle" between the sound source and the observer. Noise barrier efficiency depends principally on their design, i.e., favourable noise barriers have a diffuse element on the top (e.g., circular, Y- or T-shaped). Most important parameters which are used to describe the noise barriers are insertion loss (IL), transmission losses (TL) and barrier absorption coefficient. Usually, noise barriers can be divided into several types: Ground-mounted noise barriers (made from natural earth materials), structure-mounted noise barriers and the combination of the first two [11].

 IntechOpen

Source	Intensity	Intensity level
Threshold of hearing (TOH)	10^{-12}	0 dB
Whisper	10^{-10}	20 dB
Pianissimo	10^{-8}	40 dB
Normal conversation	10^{-6}	60 dB
Fortissimo	10^{-2}	100 dB
Threshold of pain	10	130 dB
Jet take-off	10^{2}	140 dB
Instant perforation of eardrum	10^{4}	160 dB

Figure 1.
Examples of different noise levels (from Müller, FMP, Springer 2015).

Figure 2.
Soundscape classification (example: Croatia).

When considering certain limitations of noise barriers in general, noise barriers can best serve as a solution if they are planned before the actual building (which is today a quite rare case-scenario). In addition, when incorporating a noise barrier into an existing urban environment, researchers should take into account the "visual pleasantness" and economic feasibility of the noise barrier [12, 13].

As previously mentioned, a more modern approach to noise management would be the soundscape concept. The soundscape concept modifies and complements the assessment of noise and its effects on humans [14]. Soundscape includes all the sounds from a certain acoustic environment received by human ear. These sounds can be divided into three major groups: biophony, geophony and anthrophony [14]. Soundscapes can be classified. The most common classification is the one with respect to the related environment, i.e., we can differentiate: natural soundscapes

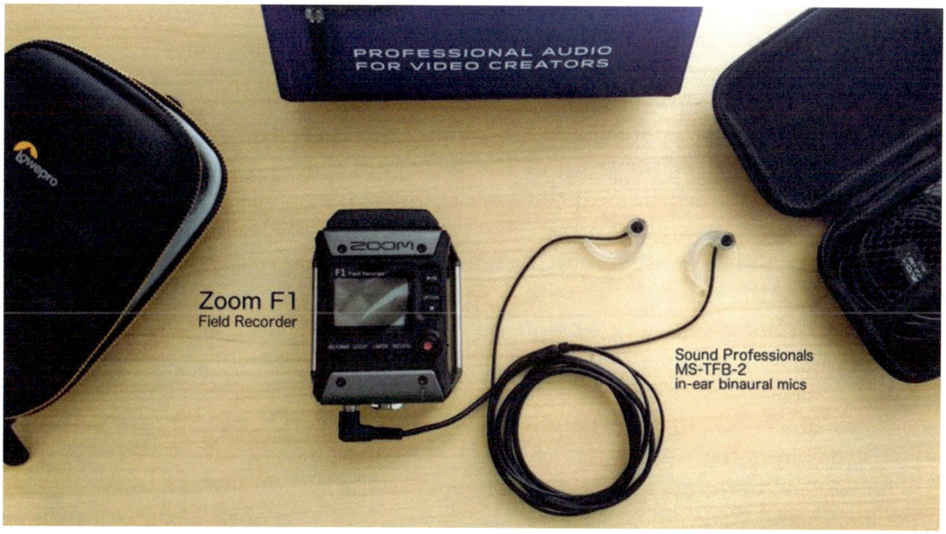

Figure 3.
Typical equipment for the soundwalk method.

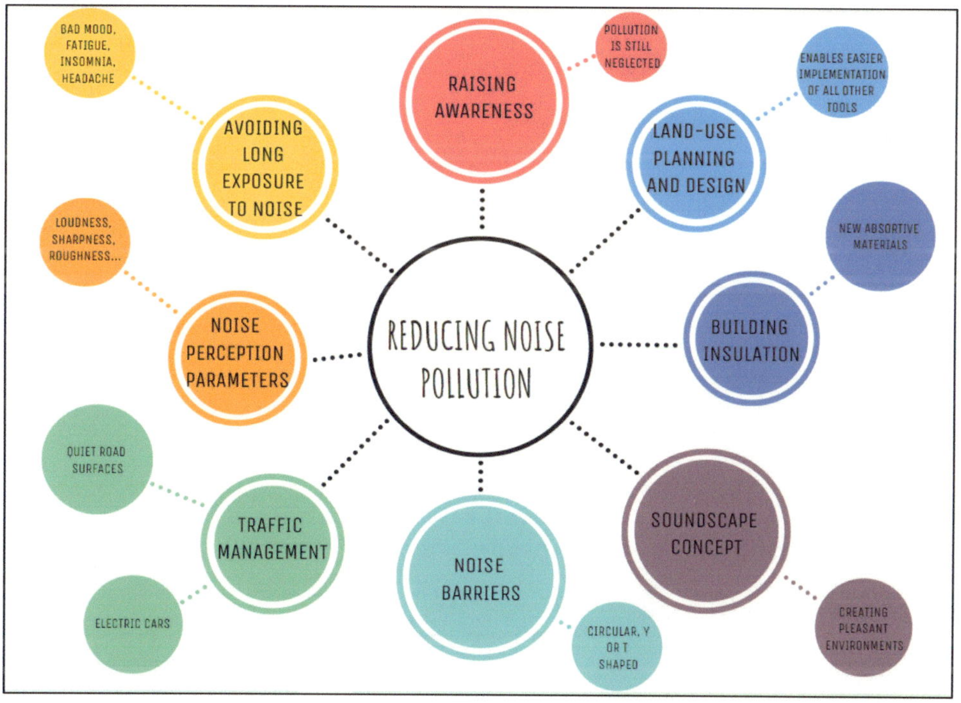

Figure 4.
Mind map of noise pollution management.

(e.g., marine, forest soundscape, etc.), rural soundscapes and urban soundscapes (**Figure 2**) [14].

Soundscapes are usually recorded using the soundwalk method which was introduced by an urban planner Kevin Lynch. The usual recording of a soundscape has the duration of 30 min. Recording takes place several times a day, for several days, however, always at a nice and dry weather. The soundwalk method uses a recorder, and a pair of binaural microphones places in the ears of the person who is performing the soundwalks, i.e., soundwalker (**Figure 3**) [15, 16].

Soundscapes are analysed in most cases using several types of questionnaires which are fulfilled by listeners or participants in studies. Possible questionnaire designs can include direct questions to listeners about the soundscape, requirements for a more detailed descriptions of the soundscape and attributes that may or may not be related to mathematical scales and adjective pairs [17–20].

Nowadays soundscape studies are oriented toward human health, well-being and overall quality of life [21–24].

Bearing in mind everything written, it can be concluded that noise pollution and its management is a very complex problem which needs an interdisciplinary approach. Experts such as urban planners, architects, doctors, biologists, psychologists as well as acoustic engineers should all collaborate and benefit from each other's work with a common cause to improve the overall quality of life. By working together, it is achievable to manage and reduce noise pollution and moreover recuperate the human health and well-being of the residents, especially the ones living and working in urban areas (**Figure 4**).

Author details

Mia Suhanek
University of Zagreb, Croatia

*Address all correspondence to: mia.suhanek@fer.hr

References

[1] World Health Organization. Guidelines for Community Noise. Geneva, Switzerland: World Health Organization; 1999

[2] Goines L, Hagler L. Noise pollution: A modern plague. Southern Medical Journal. 2007;**100**(3):287-294

[3] Ising H, Kruppa B. Health effects caused by noise: Evidence in the literature from the past 25 years. Journal Noise and Health. 2004;**6**(22):5-13

[4] Basner M, Babisch W, Davis A, et al. Auditory and non-auditory effects of noise on health. Lancet PMC. 2014;**383**(9925):1325-1332

[5] European Commission. Directorate-General for Environment. Noise impacts on health, Publications Office. 2015. DOI: 10.2779/374210

[6] Sliwi'nska-Kowalska M, Zaborowski K. WHO environmental noise guidelines for the European region: A systematic review on environmental noise and permanent hearing loss and Tinnitu. International Journal of Environmental Research and Public Health. 2017;**14**:1139

[7] De Paiva Vianna KM, Alves Cardoso MR, Calejo Rodrigues RM. Noise pollution and annoyance: An urban soundscapes study. Noise & Health. 2015;**17**:125-133

[8] Eriksson C, Pershagen G, Nilsson M. Biological Mechanisms Related to Cardiovascular and Metabolic Effects by Environmental Noise. World Health Organization; 2018

[9] Kinsler LE, Frey AR, Alan B, Coppens AB, Sanders JV. Fundamentals of Acoustics. 1982

[10] Zwicker E, Fastl H. Psychoacoustics: Facts and Models. Berlin, Heidelberg, New York: Springer - Verlag; 1999

[11] Ishizuka T, Fujiwara F. Performance of noise barriers with various edge shapes and acoustical conditions. Applied Acoustics. 2004;**65**:125-141

[12] Maffei L, Masullo M, Aletta F. Influence of the design of railway noise barriers on soundscape perception. In: Internoise. New York. 2012;**2012**:1-8

[13] Maffei L, Masullo M, Aletta F, Gabriele M. The influence of visual characteristics of barriers on railway noise perception. Science of the Total Environment. 2013;**445-446**:41-47

[14] Schafer RM. Our Sonic Environment and the Soundscape: The Tuning of the World. Destiny Books; 1994

[15] Lynch K. The Image of the City. The MIT Press; 1960

[16] Semidor C. Listening to a City with the Soundwalk method. Acta Acustica united with Acustica. 2006;**92**(6):959-964

[17] Schulte-Fortkampn B, Bennet B. Standardization in soundscape and its application. The Journal of the Acoustical Society of America. 2015;**138**(3):1748

[18] Genuit K, Bray W. Soundscape measurement and analysis. Journal of the Acoustical Society of America. 2006

[19] Kull RC. Soundscape measurements: Moving towards a standard. The Journal of the Acoustical Society of America. 2005;**118**:1875

[20] Suhanek M, Đurek I. Implementation of bipolar adjective pairs in analysis of urban acoustic environments.

Promet-Traffic & Transportation. 2016;**5**:461-470

[21] Schulte-Fortkamp B, Fiebig A. Soundscape analysis in a residential area: An evaluation of noise and people's mind. Acta Acustica United with Acustica. 2006;**92**(6):875-880

[22] Schulte-Fortkamp B, Kang J. Introduction to the special issue on soundscapes. Journal of the Acoustical Society of America. 2013;**134**(1):765-766

[23] Botteldooren D, Andringa T, Aspuru I, Brown AL, Dubois D, Guastavino C, et al. From sonic environment to soundscape. Soundscape and the Built Environment. 2015:1-26

[24] Kang J, Aletta F, Gjestland TT, Brown LA, Botteldooren D, Schulte-Fortkamp B, et al. Questions on the soundscapes of the built environment. Building and Environment. 2016:1-37

Chapter 2

Guidelines for Environmental Noise Management in Developing Countries

Dietrich Schwela

Abstract

This chapter describes the challenges of environmental noise policies and governance, noise emissions, noise transmission modeling, and those of health and economic risk assessment in developing countries. It bases on an analysis of current legislation regarding noise pollution in major developing countries in Asia, Africa, and Latin America. Although legislators are engaged in promulgating laws and regulations explicit procedures for noise measurement, noise mapping, development of a healthy and comfortable soundscape, and the implementation and enforcement of legislation are rarely developed. A strategic framework approach is needed to overcome these challenges and enable countries to achieve sustainable environmental noise management. Guidelines are provided to resolve these tasks to better protect the population of urban areas against the health end economic impacts of environmental noise.

Keywords: guiding principles, noise policies, governance, information challenges, health and economic risk assessments

1. Introduction

The World Health Organization (WHO) has considered environmental noise (also called community noise, domestic noise, or residential noise) in its environmental health criteria and guideline documents an important problem since the 1970s [1–4]. In the earliest document of 1980 noise is explicitly 'considered to be any unwanted sound that may adversely affect the health and wellbeing of individuals or populations,' and the later documents do not redefine the term. This definition is often incorrectly quoted as 'noise is unwanted sound', see for example [5, 6]. The WHO Guidelines for Community Noise define environmental noise as noise emitted from all sources, except noise at the industrial workplace [2].

Exposure to environmental noise has several impacts on human health and the environment, which have social and economic implications. These include [4, 7]:

- Cardiovascular diseases
- Increases in cardiovascular symptoms (e.g. blood pressure)

- Hearing impairment
- Cognitive effects
- Speech interference
- Sleep disturbance
- Performance deficits
- Annoyance
- Tinnitus
- Mental health effects.

The extent of the environmental noise problem is large [8]. In the European Union (EU) an estimated 113 million people are exposed to long-term day-evening-night traffic noise levels of at least 55 dB(A). 22 million people are exposed to high levels of railway noise, and 4 million to high levels of aircraft noise. Long-term exposure to environmental noise is estimated to cause 12,000 premature deaths and contribute to 48,000 new cases of ischaemic heart disease per year. 22 million people are estimated to suffer chronic high annoyance, and 6.5 million people suffer chronic high sleep disturbance. In 2011, the WHO estimated that the disability-adjusted lost life years (DALYs) due to environmental noise exposure in EU countries amounted to 60,000 years for ischaemic heart disease, 45,000 years for cognitive impairment of children, 903,000 years for sleep disturbance, 21,000 years for tinnitus, and 654,000 years for annoyance [9].

In developing countries, urbanization, industrialization, and vehicle fleet growth have increased noise emissions and imissions[1] in densely populated areas. Exposure to environmental noise significantly threatens human health and the quality of life of millions of people. Cities such as Bangkok [10], Cairo [11], Jakarta [12] and many others [13] are now having to take action to enhance their institutional and technical capabilities to estimate and control noise exposure and implement preventive actions to reduce the risks that noise poses to their citizens [14]. Data reported from 28 cities of low-and middle-income countries were found to have equivalent sound pressure levels for daytime hours of 55–91 dB(A) [13]. Night-time equivalent sound pressure levels ranged between 42 and 80 dB(A). Corresponding noise-induced impacts included high annoyance, sleep disturbance, and persistent hearing loss [13].

The degree of environmental noise exposure of urban populations is directly related to the level of society's development in a country. Societal development results in an increase in the levels of urbanization, industrialization, and transportation systems. Without appropriate intervention, environmental noise and the noise impact on communities will increase. Governments are responsible to promulgate, implement and enforce strong environmental noise strategies, policies, laws, and regulations, which are suitable to control environmental noise. Failure to do this will make it

[1] The term 'imission' is used here instead of the term 'immission' (used in the literature and pronounced 'aimission') because its pronunciation is more logical to distinguish it from the term 'emission'.

impossible to prevent a continuous increase in environmental noise pollution, and governments will be ineffective in combating it.

Mandatory noise emission and noise imission standards at the national, regional, and municipal levels are the usual instruments of a governmental 'Command and Control' approach. Regulatory standards strongly depend on a country's risk management strategy, its socio-political situation, its technical and instrumental capacities and capabilities, costs of compliance, and the existence of international agreements and guidance documents such as those of the WHO. While countries' mandatory noise emission and imission standards usually are country-specific, in general, the following issues are to be considered [2]:

- Identification of the adverse public health impacts and the population to be protected.
- The indicators for noise imission and their ranges.
- Applicable methodologies for noise monitoring, noise mapping, and noise transmission modeling.
- Procedures for testing compliance of sound pressure level indicators with noise mandatory standards.
- Standard operating procedures for control of emissions.
- Mandatory emission standards.
- Identification and implementation of responsible authorities tasked with the enforcement of regulations.
- Procurement of sufficient funding.

The Command-and-Control approach for emission and imission control at the national, regional or municipal levels strongly influences the implementation and enforcement of noise control policies. If regulatory standards are exceeded action plans to mitigate noise exposure, which address all relevant sources of noise pollution, must be drafted, implemented and enforced.

In principle, there is a need for a strategic approach (SA) on Environmental Noise Management (ENM) in developing countries to assist decision-makers and stakeholders to formulate and implement effective ENM strategies [15].

The Inter-Noise 2007 Workshop on Environmental Noise Management in Developing Countries observed [15]:

- **The importance of an overall strategy.** Although a step-by-step programme of implementation of environmental noise policies is probably the realistic way forward it should be done in the context of a clear, strategic approach. Most developing countries lack this.
- **The importance of the implementation and enforcement of environmental noise policies.** Quite a few developing countries have promulgated noise policies, but the implementation and enforcement of them are poor. This is partly the

result of a lack of political will and partly because of the cost. Because it is unrealistic to expect implementation and enforcement to rapidly improve a step-by-step approach would be more realistic.

- **The importance of active citizens' groups.** Due to poor understanding of the impacts of environmental noise among both politicians and the public the effect on stress levels, health, quality of life etc. – there is little pressure on governments from citizen groups for action to be taken. Only when these impacts are better understood will governments be motivated to tackle environmental noise and will citizens demand that noise be taken seriously. There are citizen groups in a few developing countries, however, protesting about aircraft noise and about increasing noise from traffic on existing roads.

- **The importance of low-cost solutions.** At present tackling environmental noise is not a political priority for most developing countries. Therefore, it is going to be particularly difficult to persuade them to put an effective environmental noise strategy in place if they believe it is going to cost a lot of money. Therefore, low-cost solutions are important. It also is important to highlight the cost–benefit advantages of tackling environmental noise, for example, money spent on noise reduction could result in savings on health costs.

- **The importance of not re-inventing research, policy and practice.** Developing countries can use the research that has already been done by some countries such as the United Kingdom and, more importantly, international organizations such as the WHO and the International Civil Aviation Organization (ICAO), even though many of these bodies are basing their recommendations on experience from developed countries. In fact, developing nations should get involved as they may bring a new fresh, perspective to their deliberations.

The aim of ENM is to enable government authorities to design policies and strategies to achieve and maintain a low-noise soundscape and reduce environmental noise impacts on human health and protect animals against noise exposure. To implement a low-noise soundscape, governmental authorities, in collaboration with other stakeholders, must consider the local circumstances with respect to background noise levels and the available technological and instrumental capacities and capabilities. In addition, responsible authorities must also account for extant cultural and social conditions and the financial and human resources available.

Several factors determine an effective ENM strategy. These include the knowledge of relevant environmental noise sources, the application of models for noise transmission and noise mapping, and procedures for the assessment of noise exposure and its related health and environmental impacts. The promulgation, implementation, and enforcement of emission standards and health-based imission standards are also needed as well as a range of cost-effective noise exposure control measures. Responsible authorities must be empowered to implement and enforce control measures. A simplified cycle of ENM is depicted in **Figure 1** [16, 17].

There are a lot of different stakeholders involved in ENM. These include politicians, political advisors, technology officials, policy analysts, communities, researchers, interest groups, and acoustic professionals. The interaction of these players with the policy stages involved in ENM is shown in **Figure 2**.

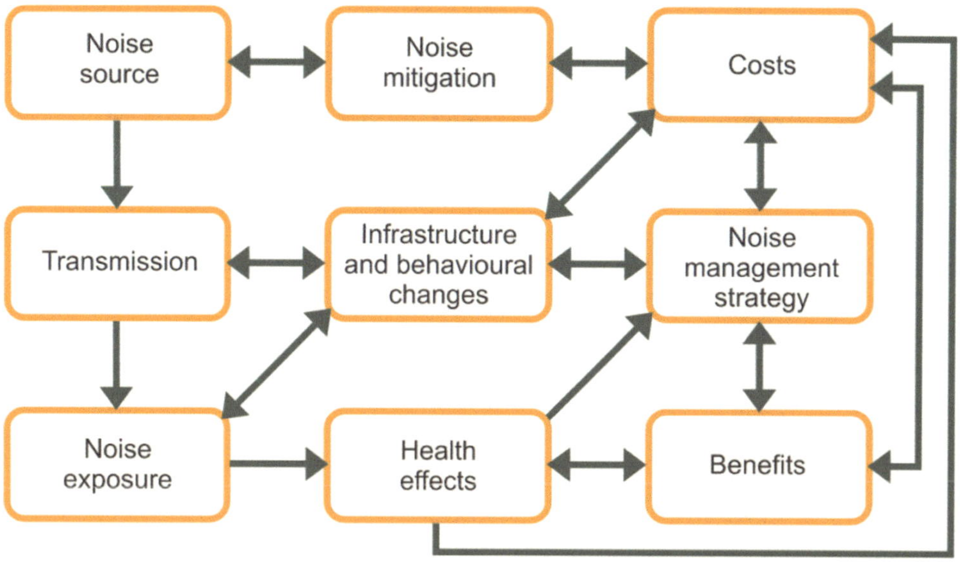

Figure 1.
A simplified cycle for environmental noise management. Source: Schwela & Finegold 2009 [16], Haq & Schwela 2012 [17].

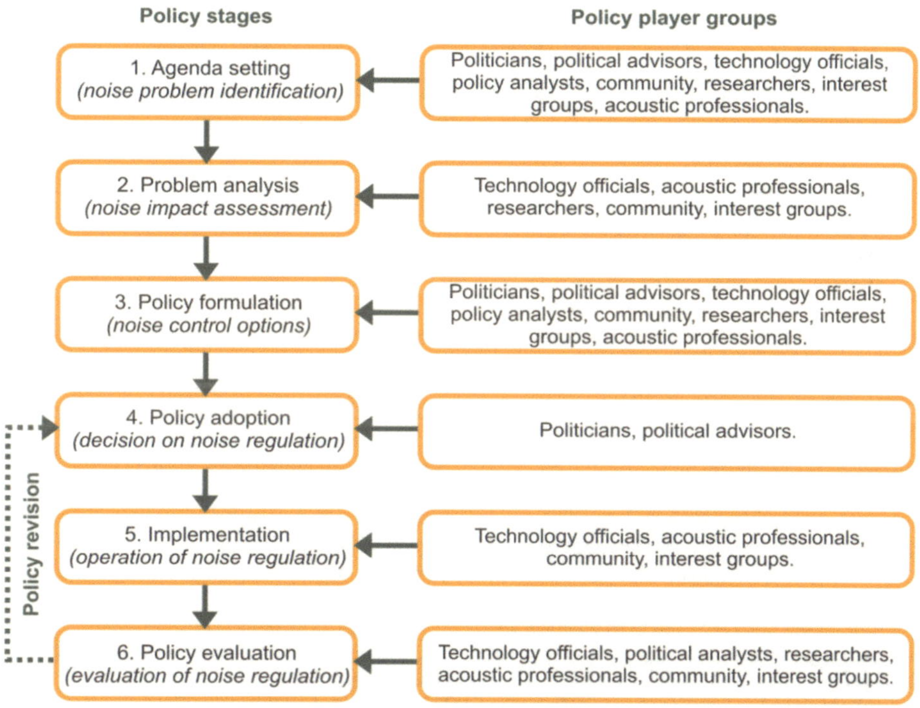

Figure 2.
Interaction between policy stages and involved stakeholders. Source: Adapted from Hede [17, 18].

A SA on ENM in developing countries systematically encompasses the most important components of comprehensive ENM. It is a flexible, rational, and broad high-level approach that is adaptable to the needs of different countries and cities. It helps guide national and local governmental authorities and other stakeholders who have a role to play in ENM. Governmental authorities in collaboration with relevant stakeholders can formulate and implement ENM strategies and programmes to prevent further deterioration of sound pressure levels. Stakeholders include the judiciary, the private sector, civil society including non-governmental organizations, the media, academia, development agencies and financial institutions.

This chapter does not develop such a SA but outlines guidelines to develop a SA and, by its realization, implementation and enforcement help reduce the health impacts of different types of environmental noise such as noise from road traffic, railways, airports and low-flying aircraft, industries, residences, leisure facilities, shooting ranges, outdoor appliances and ships in or close to ports.

2. Problem description

Guiding principles related to ENM aim to ensure the protection of human health from environmental noise. ENM should first be based on the polluter pays principle, the precautionary principle, the prevention principle, and the principle of

Access to Environmental Information: all stakeholders should have access to information regarding Noise
Awareness: Provision of information to all stakeholders
Best practice: application of state of the technology
Co-benefits: consideration of the benefits of integrated ENM, air pollution management including greenhouse gas reduction
Coherence: orientation of the efforts of all stakeholders including different neighboring jurisdictions towards a common objective.
Concerted effort: discussion and cooperation among all stakeholders involved
Compatibility: development of ENM compatible with regional, national and local needs
Continual Improvement: to promote the continual improvement of ENM as well as the reduction of noise itself
Cost-effectiveness: ENM measured at least cost and highest effectiveness
Decentralization: implementation of decentralized ENM with regional, national, and local components with due consideration to local capacity
Equity: fair and equal protection of all people from noise exposure and consideration of individual vulnerability
Integrated approach: development of integrated ENM (prevention, monitoring of adverse impacts, control of sources and education)
Opportunity: sound solutions to noise problems at the suitable moment
Participation: active participation of the population in the development and implementation of the plans to minimize noise pollution and prevent the increase of noise levels
Polluter Pays Principle: individuals responsible for noise pollution should bear the cost of its consequential impacts
Precautionary Principle: where there are threats of serious or irreversible health damage, lack of full scientific certainty should not be used as a reason for postponing cost-effective measures to prevent higher noise levels
Prevention principle: action should be taken where possible to reduce noise at the source
Stakeholder: Commitment of all stakeholders to noise management
Sustainability: development of economically and socially compatible ENM which is sustainable over the long term and future generations
Stepwise approach: ENM follows a target and milestone approach
Universality: comprehensive ENM including human health

Box 1.
The guiding principles of ENM.

participation of all stakeholders, including the population [2, 19]. The principle of participation requires the commitment of all stakeholders to ENM, their access to information regarding environmental noise, the raising of stakeholder awareness, equity with respect to the protection of the public against noise exposure, and the orientation of all stakeholders towards a common objective (coherence) in a concerted effort. Secondly, any ENM approach should be integrated with the more general efforts of environmental protection against all kinds of pollution to fully exploit the benefits of integrated solutions. Thirdly, an ENM approach should develop sound solutions that are compatible with national, regional, and local needs and, consequently, can be implemented in a decentralized way with due consideration of local capacities at least costs and highest efficiency. Fourthly, as developing countries always suffer from lack of appropriate funds, ENM is to follow a stepwise approach by setting achievable targets and milestones. Finally, ENM should be sustainable in the sense used by the WSSD report [20] and comprehensive with respect to public health protection.

However, in developing countries economic, institutional, and political constraints may hamper the full implementation of these principles.

The guiding principles are defined and summarized in **Box 1** [19].

3. Challenges in developing countries and guidelines for overcoming them

Challenges with respect to ENM exist in the fields of [13, 15–17, 21–24].

- Environmental noise policies.
- Environmental noise governance.
- Information on environmental noise emissions.
- Environmental noise modeling.
- Environmental noise monitoring.
- Health and economic risk assessments.
- Financing environmental noise management.

The following procedures to help governmental authorities in collaboration with other stakeholders to overcome these types of challenges are specified in Sections 3.1–3.6 [17, 20]:

- Identify:
    - Appropriate policies on environmental noise.
    - Relevant legislative and regulatory requirements.
    - Important sources of environmental noise caused by human activities.

- Set:
  - Appropriate objectives and targets for human (and animal) health.
  - Priorities and milestones for achieving objectives and targets.
- Establish:
  - Policies, strategies, laws, and regulations on environmental noise.
  - A structure and programmes to implement policies and achieve objectives and targets.
- Facilitate:
  - The modeling of environmental noise.
  - The estimation of effects on human health (and animals).
  - Urban planning, corrective action, and the prevention of adverse effects.
- Ensure compliance with emission and noise standards.
- Account for changing circumstances.

In addition, the barriers stated above can be overcome by:

- Gaining:
  - Ministerial support in developing countries for a rational ENM.
  - Support from international agencies, especially regarding technical and financial means needed.
- Undertaking cost–benefit analyses and health impact studies.

3.1 Challenges in developing countries in the field of environmental noise policies and guidelines to overcome them

Environmental noise policies aim at including and/or strengthening the concept of environmental noise, human (and animal) health in policies, legislation and its harmonization, implementation, and enforcement in the development of developing countries and countries in transition. As **Figure 1** and the discussion above show, several factors determine an effective and rational ENM strategy. These include noise monitoring networks, models for the transmission of sound pressure levels, noise mapping, assessments of human exposure and impacts, and the promulgation of emission standards and imission health-based standards. In addition, several cost-effective noise exposure control measures are necessary, together with the legislative powers and human and financial resources to implement and enforce them.

The following 'challenges' to achieve this aim have been reported in the context of environmental laws and politics which also analogously apply to noise legislation [17, 25, 26]:

- Low government commitment to ENM policies, their implementation and enforcement. A recent report from Kenya states that the Environmental Management and Coordination Act 'has been unable to ensure that the country fully addresses present-day environmental challenges' [27]. Similarly, a report on Nigeria's environmental governance framework has identified that it is 'ineffective in dealing with the country's environmental challenges' [28].

- Limited

    - coordination and integration of ENM policies with other sectoral policies and plans [29, 30].

    - collaboration of different responsible agencies [25].

    - institutional capacity to implement and enforce ENM legislation and policies [31].

    - control of corruption [32].

- Absence of risk-based approaches, which form a part of ENM policies and legislation [33].

- Limited appropriate review mechanisms to evaluate policies for noise mitigation measures [34].

- Absence of soundscape policies to judge exposure to a combination of noises from different sources [35].

- Lack of:

    - Criteria for guidelines/standards for compliance testing [31, 36].

    - Stakeholder participation (particularly of industry, manufacturers, urban planners, transport planners, transport associations, the informal sector, health communities, enforcement institutions and financial institutions) to formulate and implement ENM policies [25, 27].

    - A detailed cost–benefit analysis of policy measures [2].

    - Monitoring and modeling environmental noise levels [30].

    - Assessment of impact on human health and environment due to noise exposure [30].

    - Reports on sound pressure levels and their impacts in a transparent way [30].

  - o Information sharing to the public on the effects of environmental noise, raising awareness and promoting participation and engagement [30, 37].

- Use of obsolete emission and imission standards [30].

'Guidelines' for overcoming the main challenges of environmental noise policies include [16, 19]:

- The adverse impacts of environmental noise pollution on health and the environment can be mitigated, once ENM is acknowledged as an objective for sustainable development and made an integral part of the overall policy framework and is considered in specific policies such as land use planning, energy, transport, and industrial development.

- In emission and imission standard setting, social equity and fairness to all stakeholders involved (e.g. industry, local authorities, non-governmental organizations, media and the public) can be ensured if a participatory approach is followed – as far as possible and meaningful.

- In setting exposure standards and averaging times, the globally applicable WHO *Guidelines for Community Noise* [2], the WHO *Night Noise Guidelines for* Europe [3] and the WHO *Environmental Noise Guidelines for the European Region* [4] may be used.

- For the assessment of adverse health impacts due to environmental noise exposure in developing countries the WHO/EURO *Burden of Disease from Environmental Noise* [9] may give useful advice.

- Promote the inclusion of environmental noise in Environmental Impact Assessments for planned projects.

3.2 Challenges in developing countries in the field of environmental noise governance and guidelines to overcome them

The objective of environmental noise governance is to facilitate law implementation and enforcement and inform, educate, train and strengthen stakeholder participation in all aspects related to environmental noise and the prevention and reduction of environmental noise exposures and the corresponding health and environmental impacts. To achieve this objective, governmental authorities can implement the individual issues of this process in collaboration with other stakeholders. As indicated above, local circumstances with respect to background noise levels and cultural and social conditions must be considered. The estimation of the costs and benefits of ENM as well as the provision of human and financial resources are indispensable ingredients of good governance.

In developing countries 'challenges' to achieve this include:

- Conflicts through duplicated responsibilities [17].

- Introduction of inappropriate technical equipment and ignorance of its usability [38].

- Prevalence of *ad hoc* awareness raising with a focus on raising alarm [39].
- Poor information on how the public can contribute towards effective ENM [34].
- High cost of awareness-raising programmes [33].
- Design and implementation of sustainable ENM strategies are often based on incomplete knowledge [28, 40].
- Insufficiency of adequate communication strategies among stakeholders [33].
- Inadequate regulatory, planning, technical, social, institutional, and financial capacity for ENM [30].

'Guidelines' for overcoming the main challenges of environmental noise governance include [16, 19]:

- A rapid assessment of the most important sources.
- Estimation of environmental noise exposure for all the noise-sensitive areas.
- Comparison of estimated environmental noise levels with environmental noise standards.
- Identify training and capacity-enhancing needs for all stakeholders and encourage, support and promote capacity-enhancement programmes
- If capacity for public information exists it can be used to inform the public and other stakeholders on a regular basis of the importance of noise and ENM strategies and the role that the public can play in reducing emissions. If capacity for communication among stakeholders does not exist, it needs to be developed.
- A focus on 'Champions in ENM' (e.g., well-known identities and celebrities) to convey noise information, increases awareness in different public groups, and keep ENM issues high on the interest list is a very useful way to disseminate ENM information.

3.3 Challenges in developing countries in the field of information on environmental noise emissions and guidelines for overcoming them

At-source measures that reduce overall emissions are preferable to noise exposure measures. For example, for road transport reduction of sound pressure levels of the engines and tyres, traffic management and transport demand management measures are options for reducing emissions at source. For aircraft transport reducing emissions at sources is a major pillar of the Balanced Approach of the International Civil Aviation Organization [41].

This section aims to include and/or strengthen enforceable, affordable, sustainable, and highly effective measures to assess and find solutions to reduce sound emissions and, consequently reduce public exposure to adverse sound pressure levels.

The 'challenges' to achieving this objective include [17]:

- Lack of [42, 43]:
    - Emission inventories and quality-assured emission data.
    - Periodical update of emissions standards.
    - Regional harmonization of emissions standards.
    - Low-cost and effective alternative technologies.
- Short-term and ad hoc measures to reduce noise emissions usually fail to adequately address the overall challenge.
- The use of end-of-pipe solutions is not in accord with the 'precautionary' and 'prevention' principles.
- Best available control technologies are not or not consequently applied.
- Use of ineffective measures to reduce noise pollution.
- Insufficient application of the 'polluters pay' principle.
- Poor dissemination and exchange of good practices and lessons learnt (positive and negative).

'Guidelines' for overcoming main challenges of environmental noise emissions include [16, 19]:

- Replacement of short-term ad-hoc actions by medium- and long-term strategies for emission prevention and reductions will define a better way to address noise problems in developing countries and further development of these countries.
- Prevention of pollution by alternative technologies is always less expensive than a posteriori reduction of sound emissions, including the costs of health effects.
- Positive and negative lessons learnt from experiences in other countries/cities may help to rapidly find best practices and optimal solutions.
- Compilation of a (rapid) inventory of noise sources and their sound emissions is a good starting point for sound propagation estimations. A noise source inventory includes sound emissions from:
    - On-road and off-road motor vehicles.
    - Railways.
    - Airports and low-flying aircraft (i.e. close to airports).
    - Ships in or close to ports.

- Industries.
- Leisure facilities.
- Shooting ranges.
- Outdoor appliances.
- Residences.

- A periodical update (numerical reduction) of emissions standards for emitting sources and the implementation of the new standards warrants reduction in noise emissions and noise exposures. However, an emissions reduction can be traded off for an increase in the number of emitting sources (e.g. vehicles).

- Emissions standards should be regionally harmonized. Regional harmonization will support equity and help avoid the import of noisy and obsolete technology.

- Low-cost and low-noise technologies will accelerate the development of countries.

- Sound emissions from mobile sources can be reduced through a combination of measures:

    - Tighter emission standards and their enforcement.
    - Low-sound vehicle technology.
    - Inspection programmes.
    - Establishment of maintenance programmes.
    - Improved integrated land use, traffic planning and demand management on a regional scale.
    - Public transport and non-motorized transport.
    - Economic incentives/taxation.

- Emissions from stationary sources can be reduced through a combination of measures:

    - Tighter emission standards.
    - Emission control technologies and low-noise production.
    - Land use planning, zoning, and economic restructuring.
    - Enhancing enforcement.

  - Find innovative alternatives to further reduce emissions.
  - Economic incentives/taxation.

3.4 Challenges in developing countries in the field of environmental noise modeling, mapping, and monitoring and guidelines to overcome them

Environmental noise modeling has the objective to estimate national and local equivalent noise sound pressure levels in terms of L_{90}, L_{10}, L_{max}, L_{min} and L_{eq}. The result of noise modeling may be used for the development of two-dimensional (2D) and three-dimensional (3D) noise maps which provide information on noise exposure of people [44, 45]. Noise monitoring is used to assess critical sound pressure levels in residential, commercial, and industrial areas under different environmental conditions. In addition, noise monitoring can serve to validate and/or verify noise modeling predictions, and to establish and/or strengthen national and local sound pressure level monitoring programmes [46–51].

In developing countries, the 'challenges' for noise modeling and mapping are the lack of [13, 52, 53]:

- Quality-assured emission data.
- Suitable sound propagation models.
- Regional harmonization of propagation models.
- Quality-assured topographical and meteorological input data for more advanced models.

For noise monitoring, the challenges for developing countries include:

- Absence of [54]:
  - Coverage and/or limited coverage of outdoor sound pressure level monitoring systems.
  - Periodic review of sound pressure level monitoring issues.
- Limited existence of baseline data; poor quality data; lack of standard operating procedures for monitoring; poor quality control and assurance; deficiencies in the maintenance of monitoring systems; lack of monitoring of sound pressure levels in urban and peri-urban areas [55, 56].
- Insufficient representativity of monitoring sites for actual exposure of humans [57].

'Guidelines' for overcoming main challenges of environmental noise modeling, mapping, and monitoring include [16, 19]:

- Sound propagation models are useful to determine the extent and spatial coverage of noise from different sources. Propagation models can provide estimates of sound pressure levels from transportation, ports and airports, railways, and industrial plants.

- Sound pressure level monitoring mainly serves to validate the results of models and may be useful to test compliance with noise imission standards. The results of monitoring can provide feedback for continuous process of decreasing noise levels by lowering the standard values. Monitoring can also serve to better establish associations between environmental noise exposure and health impacts.

- Monitoring is usually performed at those places where people live. Hotspot monitoring may be useful for assessing exposure at locations of high noise exposure, near sources.

- Quality assurance and quality control (QA/QC) are necessary conditions to obtain reliable data (i.e. data of at least 'known quality') from a sound pressure level monitoring programme. The development of QA/QC programmes and implementation and strict obeyance of QA/QC plans to ensure that information from sound pressure level monitoring data provides a reliable basis for policy making.

- Publications on QA/QC in noise monitoring exist, which could be helpful to set up QA/QC plans and obtain data of known quality [58–60].

3.5 Challenges in developing countries in the field of health and economic risk assessments and guidelines for overcoming them

Little data exist on the human health impacts of urban noise pollution in developing countries [61]. Communities have little knowledge of impacts of noise exposure on human health, which is demonstrated by their ignorance of this threat [62]. The objective of this section is to establish and/or strengthen national and local programmes which monitor the health and economic impacts of environmental noise exposure in a harmonized way.

'Challenges' to achieving this objective include [63–67]:

- Lack of long-term studies on health due to environmental noise exposure.

- Scarcity of studies on economic impacts due to environmental noise exposure [68, 69].

- Scarcity of short-term studies on health due to environmental noise exposure [13, 35].

- Low public awareness [70–72].

- Poor information and assessment of health and economic impacts of environmental noise exposure [73].

- Low quality of evidence on noise exposure impacts [74].

- Insufficient institutional capability [75].

'Guidelines' for overcoming main challenges of health and economic risk assessments include [16, 19]:

- To protect public health and minimize the economic risk of environmental noise exposure, national and local institutions such as information and training centres should be established or strengthened, which can evaluate the health and economic impacts of environmental noise exposure.
- Cost–benefit analysis is based on reliable information on the health and environmental impacts due to environmental noise exposure. A standardized calculation of the social costs of noise exposure on human health is needed to achieve this goal. The assessment of economic and financial impacts of environmental noise exposure on human health will determine the economic costs of environmental noise exposure on society and different stakeholders. A cost–benefit analysis demonstrates the advantages of mitigating environmental noise exposure.
- It is advantageous to train and educate administrative staff and general professionals on the topic of environmental noise-induced health effects.

3.6 Challenges in developing countries in the field of financing of ENM and guidelines for overcoming them

Funds are required to:

- Update and upgrade existing laws and regulations.
- Strengthen governmental institutions to implement and enforce mandatory emission and imission standards.
- Enhance capacities and capabilities for assessing source sound pressure emissions.
- Raise awareness of all relevant stakeholders.
- Promote participation and engagement of stakeholders including the public.
- Assess data of known quality of noise exposure by means of monitoring and modeling sound pressure levels.
- Produce noise maps.
- Test compliance with imission standards.
- Assess noise impacts on health and economic costs.
- If imission standards are not met, develop emission control measures, implement, and enforce them.

This section discusses challenges and provides guidelines to establish mechanisms for financial sustainability in regional, national, and local noise, health programmes including financing from the private sector and other sectors.

Challenges' in developing countries include [17]:

- Environmental noise pollution currently is considered an issue of low priority and, therefore, underfunded.
- Eventually, existing resources are inefficiently used.
- There is a lack of:
    - Good governance regarding financing with a high level of accountability and transparency.
    - Sufficient funding for institutional capacity enhancement.
    - Knowledge of opportunities in applying existing market mechanisms.
    - Co-operation and coordination among funding agencies.
    - Implementation and enforcement of the 'polluter pays principle'.

'Guidelines' for overcoming main challenges of financing of ENM include [16, 19]:

- Raising awareness among decision-makers on the need for financing ENM and the monitoring of noise-induced impact on health is crucial.
- Governments could share information on ENM and give incentives to the private sector to participate in ENM according to the polluter pays principle.
- Economic, financial, and cost efficiency/benefit analyses for ENM including health programmes would constitute a clear procedure to limit expenditures for public health impacts.
- International aid agencies could be helpful in capacity enhancement to reduce noise exposure as an impediment to development. These agencies and regional and national funding institutions could provide incentives for ENM.
- To maximize synergies, it would be useful to coordinate funding among governmental agencies.

4. Conclusions

Environmental noise pollution is growing in developing countries because of an increase in the levels of urbanization, industrialization, and transportation systems. The degree of environmental noise exposure of urban populations is directly related to the level of society's development. Environmental noise and the noise impact on communities will increase if appropriate interventions are not considered. It is the responsibility of Governments to promulgate, implement and enforce strong environmental noise strategies, policies, laws, and regulations, which are suitable to control of environmental noise. If Governments fail to do so, it will be impossible to prevent a continuous increase in environmental noise pollution, and they will be ineffective to develop an environmental noise management system. The aim of environmental noise management is to maintain low-noise soundscapes that protect human and animal health.

This chapter reviews the challenges in developing countries with respect to environmental noise policies, governance, noise emission, noise modeling, mapping, and monitoring, the assessment of health and economic risks, and the mechanisms for financing environmental noise management. Guidelines are presented to create a strategic approach for environmental noise management suitable for developing countries that will help overcome these challenges.

Acknowledgements

The author thanks Lawrence Finegold, Research Psychologist at Finegold & So, Consultants, Dayton, Ohio, United States, and Gary Haq, Stockholm Environment Institute, University of York, Heslington, York, United Kingdom, for their valuable discussions on and contribution to the strategic framework for environmental noise management.

Conflict of interest

The author declares no conflict of interest.

Author details

Dietrich Schwela
Department of Environment and Geography, Stockholm Environment Institute, University of York, York, United Kingdom

*Address all correspondence to: dietrich.schwela@york.ac.uk

References

[1] WHO. Noise. Environmental Health Criteria Document No. 12. Geneva: World Health Organization. International Programme on Chemical Safety; 1980 In conjunction with the United Nations Environmental Programme. Available from: http://www.inchem.org/documents/ehc/ehc/ehc012.htm

[2] WHO. In: Berglund B, Lindvall T, Schwela DH, editors. Guidelines for Community Noise. Geneva: World Health Organization; 1999. Available from: https://apps.who.int/iris/handle/10665/66217

[3] WHO. Night Noise Guidelines for Europe. Copenhagen: WHO Regional Office for Europe; 2009. Available from: http://www.euro.who.int/__data/assets/pdf_file/0017/43316/E92845.pdf

[4] WHO. Environmental Noise Guidelines for the European Region. Copenhagen: WHO Regional Office for Europe; 2018. Available from: http://www.euro.who.int/en/publications/abstracts/environmental-noise-guidelines-for-the-european-region-2018

[5] Basner M, Clark C, Hansell A, et al. Aviation noise impacts: State of the science. Noise & Health. 2017;**19**(87):41-50. Available from: https://www.ncbi.nlm.nih.gov/pmc/articles/PMC5437751/

[6] Fink D. A new definition of noise: Noise is unwanted and/or harmful sound. Noise is the new 'secondhand smoke'. In: Acoustical Society of America. 178th Meeting. 2-6 December 2019. San Diego, CA, USA: Acoustical Society of America; 2019. Available from: https://asa.scitation.org/doi/pdf/10.1121/2.0001186

[7] APHA. Noise as a Public Health Hazard. Policy Statement 2021. Washington, DC, USA: American Public Health Association; 2021. Available from: https://apha.org/Policies-and-Advocacy/Public-Health-Policy-Statements/Policy-Database/2022/01/07/Noise-as-a-Public-Health-Hazard

[8] EEA. Environmental noise in Europe – 2020. Copenhagen: European Environment Agency; 2020. Available from: https://www.eea.europa.eu/publications/environmental-noise-in-europe

[9] WHO. Burden of Disease from Environmental Noise. Copenhagen: WHO Regional Office for Europe; 2011. Available from: http://www.euro.who.int/__data/assets/pdf_file/0008/136466/e94888.pdf

[10] PCD. Noise Monitoring Report. September 14, 2022, 1:00 am. Bangkok: Pollution Control Department; 2022. Available from: http://noisemonitor.net/web/

[11] Sakr A, El-Dars F, Hewehey M. Monitoring noise levels along the greater Cairo region ring road in Egypt. Journal of Environmental Sciences. 2020;**49**(2):1-25. Available from: https://journals.ekb.eg/article_158105.html

[12] Prasetyo S, Kusnoputranto H, Alikorda HS, Koestoer RH. Model of noise propagation in urban area: A case study in Jakarta. OIDA International Journal of Sustainable Development. 2016;**9**(2):45-50. Available from: https://papers.ssrn.com/sol3/papers.cfm?abstract_id=2739810

[13] Schwela D. Environmental noise challenges and policies in low and middle- income countries. SFJH. 2020;**2**(1):26-45. DOI: 10.46981/sfjhv2n1-003

[14] UNEP. Frontiers 2022: Noise, Blazes, and Mismatches. Chapter 1: Listening to Cities. From Noisy Environments to Positive Soundscapes. Nairobi: United Nations Environment Programme; 2022. Available from: https://www.unep.org/resources/frontiers-2022-noise-blazes-and-mismatches

[15] Schwela D, Finegold L, Stewart A. A strategic approach on environmental noise management in developing countries. In: 9th International Congress on Noise as a Public Health Problem. International Commission on the Biological Effects of Noise (ICBEN). Mashantucket: Foxwoods, CT; 2008. Available from: http://www.icben.org/2008/pdfs/schwela_et_al.pdf

[16] Schwela DH, Finegold LS. A strategic approach for environmental noise management in developing countries. In: Paper Presented at the Inter-Noise 2009 Conference "Innovation in Practical Noise Control, August 23–26, 2009". Ottawa, Canada; Indianapolis, IN: Institute of Noise Control Engineering of the USA; 2009. Available from: https://www.proceedings.com/06792.html

[17] Haq G, Schwela D. Transport and Environment in Sub-Saharan Africa. TEST - Transport and Environment Science Technology Network. Stockholm Environment Institute, Department of Environment & Geography, University of York; York, UK: Stockholm Environment Institute at the University of York; 2012. Available from: https://mediamanager.sei.org/documents/Publications/SEI-Pub2197-Haq-Schwela-test_tande_v2.pdf

[18] Hede A. Environmental noise regulation: A public policy perspective. In: Carter NL, Job RFS, editors. Noise as a Public Health Problem (Noise Effects '98). Vol. 2. Sydney, Australia: Noise Effects "98 PTY Ltd; 1998. pp. 687-696

[19] Schwela D, Finegold L. A strategic Approach on Environmental Noise Management in Developing Countries. Presentation at the 12th ICBEN Congress on Noise as a Public Health Problem, 18-22 June 2017. Zurich, Switzerland: Eidgenössische Technische Hochschule; 2022. Available from: https://organizers-congress.org/frontend/index.php?page_id=240320

[20] Report on the World Summit on Sustainable Development, Johannesburg, South Africa, 26 August - 4 September 2002. New York: United Nations Digital Library. 2002. Available from: https://digitallibrary.un.org/record/478154

[21] Schwela D, Finegold L, Stewart A. A strategic approach on environmental noise management in developing countries. 9th International Congress on Noise as a Public Health Problem. International Commission on the Biological Effects of Noise (ICBEN), 21-25 July 2008, Foxwoods, CT. Washington, DC, USA: American Speech-Language-Hearing Association; 2008. Available from: http://www.icben.org/2008/pdfs/schwela_et_al.pdf

[22] Howes M, Wortley L, Potts R, Dedekorkut-Howes A, Serrao-Neumann, Davidson J, et al. Environmental sustainability: A case of policy implementation failure? Sustainability. 2017;**9**(2):165. DOI: 10.3390/su9020165. Available from: https://www.mdpi.com/2071-1050/9/2/165/pdf

[23] Greenspan Bell R, Russell C. Environmental policy for developing countries. Issues in Science and Technology. 2002;**18**(3). Available from: https://issues.org/greenspan/

[24] Yuen FK. A vision of the environmental and occupational noise pollution in Malaysia. Noise & Health. 2014;**16**(73):427-436. DOI: 10.4103/1463-1741.144429. Available from: http://www.noiseandhealth.org/article.asp?issn=1463-1741;year=2014;volume=16;issue=73;spage=427;epage=436;aulast=Yuen

[25] Sánchez-Ocampo EM, Téllez-López AM, Islas-Flores H, Gómez-Oliván LM. Environmental laws and politics, the relevance of implementing regulation of the presence of emerging pollutants in Mexico: A systematic review. Water Emerging Contaminants & Nanoplastics. 2022;**1**:6. DOI: 10.20517/wecn.2021.04. Available from: https://www.oaepublish.com/wecn/article/view/4690

[26] Lau MHY, Leung KMY, Wong SWY, Wang H, Yan Z-G. Environmental policy, legislation, and management of persistent organic pollutants (POPs) in China. Environmental Pollution. 2012;**165**:182-192. Abstract: https://pubmed.ncbi.nlm.nih.gov/22285063/

[27] Wambua C. Environmental Law 2021. Trends and Development. Last updated November 30, 2021. London, UK: Chambers and Partners; 2021. Available from: https://practiceguides.chambers.com/practice-guides/environmental-law-2021/kenya/trends-and-developments

[28] Ogunkan DV. Achieving Sustainable Environmental Governance in Nigeria: A Review for Policy Consideration. Urban Governance. 2022;**2**(1):212-220. Available from: https://www.sciencedirect.com/science/article/pii/S2664328622000213

[29] Bahauddin KM. Environmental system management and governance needs in a developing country. Environment Systems and Decisions. 2014;**34**:342-357 Abstract: https://link.springer.com/article/10.1007/s10669-013-9472-3

[30] Krasnova IO, Vlasenko VN. Strategic regulatory instruments in environmental law of Russia. Journal of Siberian Federal University. Humanities & Social Sciences. 2020;**13**(10):1671-1678. Available from: http://journal.sfu-kras.ru/en/article/137547

[31] Francesch-Huidobro M, Lo CW-H, Tang S-Y. The local environmental regulatory regime in China: Changes in pro-environment orientation, institutional capacity, and external political support in Guangzhou. Environ Plan A: Economy and Space. 2012;**44**(10):2493-2511 Abstract: https://journals.sagepub.com/doi/10.1068/a44504

[32] Swain RB, Kambhampati US, Karimu A. Regulation, governance, and the role of the informal sector in influencing environmental quality? Ecological Economics. 2020;**173**:106649. Abstract: https://www.sciencedirect.com/science/article/abs/pii/S0921800919315861

[33] Hellmuth T, Classen T, Kim R, Kephalopoulos S. Methodological guidance for estimating the burden of disease from environmental noise. Copenhagen, Denmark: WHO Regional Office for Europe. Ispra, Italy: European Commission Joint Research Centre; 2012. Available from: https://www.euro.who.int/__data/assets/pdf_file/0003/177816/Methodological-guidance-for-estimating-the-burden-of-disease-from-environmental-noise.pdf

[34] Lázaro B. Comparative Study on the Institutionalisation of Evaluation in Europe and Latin America. Eurosocial - Programme for Social Cohesion in Latin America; 2015. Madrid, Spain: EUROsociAL Programme; 2015.

Available from: http://sia.eurosocial-ii.eu/files/docs/1456851768-E_15_ENfin.pdf

[35] Bouzier TAK, Berkouk D, Schwela D, Lahlouh M. Noise pollution policies in the Arab world: An overview and comparison with EU legislation. Paper submitted to Acoustics Australia. 2022

[36] Ministry of Environment and Forests 2000. The Noise Pollution (Regulation and Control) Rules. 2000. (As amended till 10/08/2017 vide S.O. 2555(E)). Available from: http://dstegoa.gov.in/Noise%20rules-10-08-2017.pdf

[37] Flacke J, Aguilar R, Pfeffer K. Citizen participation in noise action planning: Co-designing an open and interactive geospatial tool. Contribution s1-s13. 6th International Conference URBAN E-PLANNING 2022, 7.-12.4.2022. Lisbon, Portugal. Enschede, Netherlands: University of Twente. 2022. Available from: https://research.utwente.nl/en/publications/citizen-participation-in-noise-action-planning-co-designing-an-op

[38] Schwela D, Haq G. A Strategic Framework for Air Quality Management in Asia. York, UK: Stockholm Environment Institute at the University of York; 2004. Available from: https://www.researchgate.net/publication/263970515_A_Strategic_Framework_for_Air_Quality_Management_in_Asia

[39] Schwela D. Review of Urban Air Quality in Sub-Saharan Africa Region - Air Quality Profile of SSA Countries. Report 67794. Washington, DC, USA: World Bank; 2012. Available from: https://openknowledge.worldbank.org/bitstream/handle/10986/26864/677940WP0P07690020120Box367897B0ACS.pdf

[40] Grunwald A. Working towards sustainable development in the face of uncertainty and incomplete knowledge. Journal of Environmental Policy & Planning. Governance for sustainable development: coping with ambivalence, uncertainty, and distributed power. 2007;**9**(3–4):245-262. Abstract: https://www.tandfonline.com/doi/abs/10.1080/15239080701622774?journalCode=cjoe20

[41] ICAO. White paper on aircraft noise aviation noise impacts: State of the science. In: Basner M, Clark C, Hansell A, et al., editors. ICAO 2016 Environmental Report. Toronto: International Civil Aviation Organization; 2016. pp. 30-37. Available from: https://www.icao.int/environmental-protection/pages/env2016.aspx

[42] Zambrano-Montserrate MA, Ruano MA. Does environmental noise affect housing rental prices in developing countries? Evidence from Ecuador. Land Use Policy. 2020;**87**: 104059 Abstract and section snippets: https://www.sciencedirect.com/science/article/abs/pii/S0264837719304247

[43] United Nations Conference on Trade and Development 2021. The Least Developed Countries in the Post-COVID World: Learning from 50 Years of Experience. The LDC Report 2021. Available from: https://unctad.org/system/files/official-document/ldc2021_en.pdf.

[44] Abdur-Rouf K, Shaaban K. Measuring, mapping, and evaluating daytime traffic noise levels at urban road intersections in Doha, Qatar. Future Transportation. 2022;**2**:625-643. Available from: https://www.mdpi.com/2673-7590/2/3/34

[45] Akhtar N, Ahmad K, Alam P. Noise monitoring and mapping for some pre-selected locations of New Delhi, India. Fluctuation and Noise Letters. 2016;**15**(2):

1650019. Available from: https://www.researchgate.net/publication/303977013_Noise_Monitoring_and_Mapping_for_Some_Pre-selected_Locations_of_New_Delhi_India

[46] Gabrielian AB, Puranik TG, Bendarkar M, et al. Noise model validation using real world operations data. In: AIAA Aviation 2021 Forum. August 2–6, 2021. Arlington, VA, USA: Aerospace Industries Association of America; 2021. DOI: 10.2514/6.2021-2136

[47] Schlatter F, Piquerez A, Habermacher M, et al. Validation of large-scale noise exposure modelling by long-term measurements. Noise Mapping. 2017; **4**(1):75-86. DOI: 10.1515/noise-2017-0006

[48] Kushwaha S, Ahmad SK, Garg SK. Noise monitoring, modelling, and model validation at the vicinity of Noida city Centre. IRJET. 2020;7(10):1107-1111. Available from: https://www.irjet.net/archives/V7/i10/IRJET-V7I10193.pdf

[49] Bąkowski A, Dekýš V, Radziszewski L, Skrobacki Z. Validation of traffic noise models. In: AIP Conference Proceedings 2077. Montréal, Canada: Aeronautical Information Publication; 2019. p. 020005-1-020005-9. Available from: https://aip.scitation.org/doi/abs/10.1063/1.5091866

[50] Jacyna M, Wasiak M, Lewzuk K, Karoń G. Noise and environmental pollution from transport: Decisive problems in developing ecologically efficient transport systems. Journal of Vibroengineering. 2017;**19**(7):5639-5655. DOI: 10.21595/jve.2017.19371

[51] Onjefu SA, Ejembi E, Onjefu LA. Measurement of noise pollution in northern industrial areas of Windhoek, Namibia. Journal of Environmental Protection. 2019;**10**:1144-1154. Available from: https://www.scirp.org/journal/jep

[52] Clark SN, Alli AS, Ezzati M, et al. Spatial modelling and inequalities of environmental noise in Accra, Ghana. Environmental Research. 2022;**214**(2): 113932. DOI: 10.1016/j.envres.2022.113932

[53] Meller G, de Lourenço WM, de Melo VS, de Campos GG. Use of noise prediction models for road noise mapping in locations that do not have a standardized model: A systematic review. Preprint from Research Square. 2022;**2022**. DOI: 10.21203/rs.3.rs-2121992/v1

[54] Kalisa E, Irankunda E, Rugengamanzi E. Amani M. noise levels associated with urban land use types in Kigali, Rwanda. Heliyon. 2022;**8**:e10653. Available from: https://www.cell.com/heliyon/pdf/S2405-8440(22)01941-7.pdf

[55] Alam P, Ahmad K, Afsar SS, Akhtar N. Noise monitoring, mapping, and modelling studies – A review. JEE. 2020;**21**(4):82-93. DOI: 10.12911/22998993/119804.

[56] Ranpise RB, Tandel BN. Urban road traffic noise monitoring, mapping, modelling, and mitigation: A thematic review. Noise Mapping. 2022;**9**(1):48-66. DOI: 10.1515/noise-2022-0004

[57] Olopade A. Effect of noise pollution on hearing in Nigeria. In: The Guardian. Earcare Foundation. London, UK: Guardian Media Group; 2021. p. 2021. Available from: https://guardian.ng/opinion/effect-of-noise-pollution-on-hearing-in-nigeria/

[58] Ibekwe TS, Folorunsho DO, Dahilo EA, et al. Evaluation of mobile smartphones app as a screening tool for environmental noise monitoring. Journal of Occupational and Environmental Hygiene. 2016;**13**(2):D31-D36. DOI: 10.1080/15459624.2015.1093134

[59] MEE. Technical Specifications for Environmental Noise Monitoring Routine Monitoring for Urban Environmental Noise. Beijing: Ministry of Ecology and Environment; 2013 In Chinese. Available from: https://english.mee.gov.cn/Resources/standards/Noise/Method_standard3/201304/t20130410_250575.shtml

[60] Maijala P, Shuyang Z, Hetttola T, Virtanen T. Environmental noise monitoring using source classification in sensors. Applied Acoustics. 2018;**129**(1):258-267. DOI: 10.1016/j.apacoust.2017.08.006

[61] HEI. Urban air and noise pollution in sub-Saharan Africa: A study of prenatal exposures, birth outcomes, and sleep disturbances in infants. In: Poster Abstract for HEI Annual Conference 2022. Boston: Health Effects Institute; 2022. Available from: https://www.healtheffects.org/research/ongoing-research/urban-air-and-noise-pollution-sub-saharan-africa-study-prenatal-exposures

[62] Baffoe PE, Duker AA, Senkyire-Kwarteng EV. Assessment of health impacts of noise pollution in the Tarkwa Mining Community of Ghana using noise mapping techniques. Global Health Journal. 2022;**6**(1):19-29. Available from: https://www.sciencedirect.com/science/article/pii/S2414644722000069

[63] Ana GREE, Shendell DG, Brown GE, Sridhar MKC. Assessment of noise and associated health impacts at selected secondary schools in Ibadan, Nigeria. JEPH. 2009;**209**:739502. Available from: https://www.hindawi.com/journals/jeph/2009/739502/

[64] Effah E. Health effects caused by noise - the case of Africa: Evidence in literature from the past 25 years. Asian Journal of Advanced Research and Reports. 2022;**16**(2):19-27. DOI: 10.9734/AJARR/2022/v16i230452

[65] Fuente A, Hickson L. Noise-induced hearing loss in Asia. International Journal of Audiology. 2011;**50**(sup 1):S3-S10. DOI: 10.3109/14992027.2010.540584

[66] Knight C, Richards GC. Noise pollution. Oxford Healthy Cities Toolkit; Oxford, UK: Kellogg College 24; 2022. Available from: https://www.healthycitiescommission.org/toolkit/noise-pollution/

[67] Gilani TA, Mir MS. Association of road traffic noise exposure and prevalence of coronary artery disease: A cross-sectional study in North India. Environmental Science and Pollution Research. 2021;**28**:53458-53477. DOI: 10.1007/s11356-021-14582-2

[68] Zambrano-Monserrate MA, Ruano M. Does environmental noise affect housing rental prices in developing countries? Evidence from Ecuador. Land Use Policy, 2019;**87**(9):104059. Available from: https://www.sciencedirect.com/science/article/abs/pii/S0264837719304247

[69] Egbenta IR, Uchegbu SN, Ubani E, Akalemeaku OJ. Effects of noise pollution on residential property value in Enugu urban, Nigeria. SAGE Open. 2021;**11**(3). DOI: 10.1177/21582440211032167

[70] Jamir L, Nongkynrih B, Gupta SK. Community noise pollution in urban India: Need for public health action. Indian Journal of Community Medicine. 2014;**39**(1):8-12. Available from: https://www.ncbi.nlm.nih.gov/pmc/articles/PMC3968587/

[71] Siddiqui IA, Nizami S, Chandio RR, et al. Consequences of traffic noise in

residents of Karachi, Pakistan. Pakistan Journal of Medical Sciences. 2015;**31**(2): 448-452. Available from: https://www.ncbi.nlm.nih.gov/pmc/articles/PMC4476360/

[72] Gilani TA, Mir MS. A study on the assessment of traffic noise induced annoyance and awareness levels about the potential health effects among residents living around a noise-sensitive area. Environmental Science and Pollution Research. 2021;**28**:63045-63064. DOI: 10.1007/s11356-021-15208-3

[73] Orikpete OF, Leton TG, Momoh OLY, Okwu MO. Appraisal of industrial and environmental noise regulation in Nigeria and its impact on sustainable national development. International Journal of Science Technology Research. 2021;**10**(9):92-103. Available from: https://www.ijstr.org/final-print/sep2021/Appraisal-Of-Industrial-And-Environmental-Noise-Regulation-In-Nigeria-And-Its-Impact-On-Sustainable-National-Development.pdf

[74] Rahman ZU, Ahmad I, Ditta A, et al. Types, sources, socioeconomic impacts, and control strategies of environmental noise: A review. Environmental Science and Pollution Research. 2022;**29**: 81087-81111. DOI: 10.1007/s11356-022-23328-7

[75] Ortmann S. Environmental Governance in Vietnam. Institutional Reforms and Failure. Department of Asian and International Studies. Kowloon Tong, Hong Kong: City University of Hong Kong; 2017. Available from: http://www.viet-studies.net/kinhte/EnvironmentReformsVN_2017.pdf

Chapter 3

Methodological Aspects of Industrial and Transport Noise Monitoring

Sergey Dragan and Aleksey Bogomolov

Abstract

The chapter outlines the methodological aspects of monitoring industrial and transport noise, including the main physical characteristics, features of sources, measuring instruments, features of hygienic regulation of industrial and transport noise, means and methods of protection against it. It is shown that industrial facilities and most modes of transport are sources of high-intensity noise, the spectrum of which is dominated by frequencies of the low-frequency infrasonic range. The close physical nature of these ranges contributes to the propagation of such noise with low attenuation, and they have good penetrating power, so most noise protection devices are ineffective. This requires careful medical supervision of persons working in such conditions, improvement of means and methods of protection against industrial and transport noise.

Keywords: transport noise monitoring, industrial noise monitoring, low-frequency noise, infrasound, hygienic noise regulation, noise protection

1. Introduction

In accordance with modern concepts, noise and infrasound are classified as harmful and dangerous physical factors, the impact of which causes a decrease in efficiency and reliability of activity, and a long or short cumulative effect causes the development of a number of diseases [1–3].

To date, a large amount of data has been accumulated on the adverse effects of noise on humans. The nature of this influence depends on the sound level, the duration of exposure and the spectral composition of the noise [4–6]. The hearing organ is the critical organ of the body when exposed to noise. It is generally accepted that the most harmful effect on the organ of hearing is provided by noise, the spectrum of which is dominated by high frequencies of the sound range (from 1 to 8 kHz). In the clinical picture, along with hearing impairment, pathology of the cardiovascular and nervous systems is often found, which made it possible to form the concept of "noise disease" [7].

The physical characteristics of infrasound are well studied by acousticians, however, hygienists and occupational pathologists have long been limited in their research by the lack of reliable and affordable measuring equipment. Therefore, the history of the study of infrasound as a factor in the environment and production environment is relatively short. The first publications devoted to the action of infrasound appeared in the period 1970–1980. During this period, reports appeared in the scientific literature about the high biological efficiency of infrasound [1, 7, 8]. The first hygienic standards for infrasound in the USSR were adopted only in 1981, while the first noise standards for its limitation in workplaces were adopted in 1956. Subsequently, a large number of publications appeared, which reflect the point of view of hygienists on the problem of infrasound effects on humans [1, 7].

Since 2004, infrasound has been included in the list of harmful and hazardous production factors in Russia. The critical organs under the influence of infrasound include not only the organ of hearing, but also the vestibular analyzer, the central and autonomic nervous system, the circulatory and respiratory organs [9–12]. The presence of several organs and systems in the clinical picture allows us to speak about the separation of infrasound pathology into a separate nosological form [1].

There are reports that low-frequency noise can have a harmful effect not only on the organ of hearing, but also on other human organs and systems. Its biological effect has a certain similarity with the effect of infrasound on the human body. An analysis of industrial and transport noise shows that its spectrum is dominated by low frequencies of the audible and infrasound ranges. Close physical parameters and biological effects allowed a number of authors to introduce the terms "low-frequency acoustic oscillations", "infrasonic disease" and "vibroacoustic disease" [1, 7].

2. Specificity of industrial and transport noise monitoring

2.1 Physical characteristics

Depending on the frequency, acoustic vibrations are divided into infrasonic, sonic and ultrasonic. According to their physical nature, the acoustic vibrations of these ranges are the same, and their separation is somewhat arbitrary and is associated with the physiological feature of the human auditory analyzer. It is believed that a person hears sounds with frequencies from 16 Hz to 20 kHz. The area of sound frequencies or acoustic vibrations of the air in the infra-, ultra-, and hypersonic ranges is not perceived by the human ear. It should be noted that modern regulatory documents give a slightly different frequency gradation for infrasound. Infrasound is commonly understood as acoustic vibrations with a frequency below 22 Hz. At high levels of sound pressure (SPL) infrasound (over 120 dB), a person has a feeling of pulsation, pressure, and even pain in the eardrum. The physical features of infrasound include a long wavelength and low absorption in the atmosphere and the resulting ability of infrasound to propagate over long distances from the source without significant loss of energy. It should be kept in mind that sound propagates spherically and the decrease in sound pressure is inversely proportional to the square of the distance from the source [13, 14].

The audio frequency range includes acoustic vibrations from 20 Hz to 20 kHz, which are perceived by the human ear. Noise is a disorderly combination of sounds of different strength and frequency. According to the predominance of acoustic energy in one or another part of the spectrum, noise is divided into low-frequency (up to 500 Hz), medium-frequency (from 500 to 1000 Hz) and high-frequency (from 1000 to 8000 Hz).

2.2 Industrial sources of noise and infrasound

The noise generated during the operation of modern production equipment, the operation of machinery and vehicles, is acoustic vibrations in a wide frequency spectrum: from infrasonic to ultrasonic ranges.

The use of various mechanisms and machines in production activities, an increase in their power and dimensions have led to a change for the worse in the acoustic situation at the workplaces of personnel. There is a tendency to increase the contribution of low-frequency components, including infrasound, to the industrial noise spectrum. Production low-frequency noise and infrasound are generated during the cyclic movement of large surfaces, during shock excitation of structures, reciprocating and rotational movement of large masses with a repetition rate of cycles of no more than 20 per second, with the rapid movement of large volumes of liquid and air. In "pure" form, infrasound practically does not occur in a production environment: as a rule, its "companions" are high-intensity noise and general vibration [1].

The spectra of most industrial and transport noises contain low-frequency noise and high-level infrasound. The results of acoustic measurements show that if airborne noise levels are about 90–100 dBA, then the presence of infrasound with a SPL of 100–107 dB can be expected [15].

Acoustic measurements at the enterprises of the metallurgical industry near blast furnaces and steel-smelting furnaces showed the presence of SPL of 95–108 dB at frequencies of 8–31.5 Hz. In the gas and oil industry, sources of low-frequency noise and infrasound are air and reciprocating compressors, ventilation installations, pipelines, and so on. SPLs from 92 to 123 dB in octave bands of 8–63 Hz were registered at workplaces. The maximum SPL in octaves of 4–31.5 Hz during the operation of ventilation units and air conditioning systems is 98–100 dB, during the operation of compressor units – 92–123 dB at frequencies of 8–16 Hz and diesel units 111–123 dB at frequencies of 8– 63 Hz (**Figures 1–4**) [1, 7, 8].

There are a large number of noise sources in the aviation industry, especially at the stages of testing individual units and components and engines. At the workplaces of aviation specialists, SPLs reach 132 dB in the high-frequency and mid-frequency

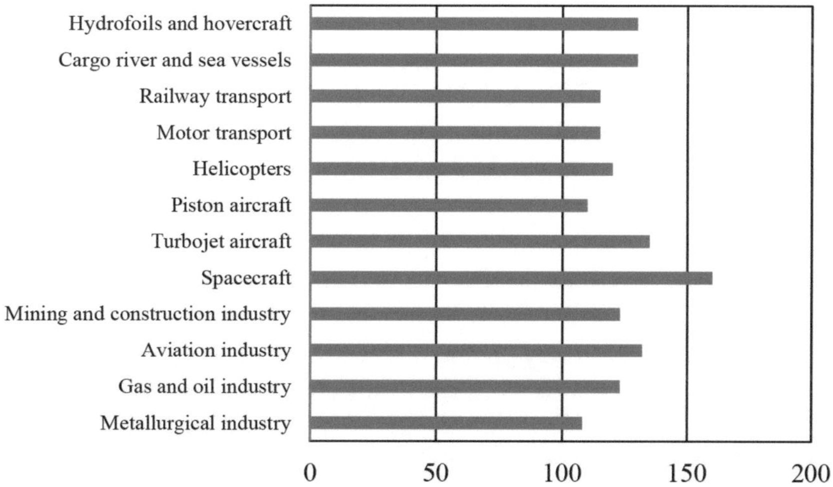

Figure 1.
Sources of low-frequency noise and infrasound in industry and transport (abscissa—Sound pressure levels, dB).

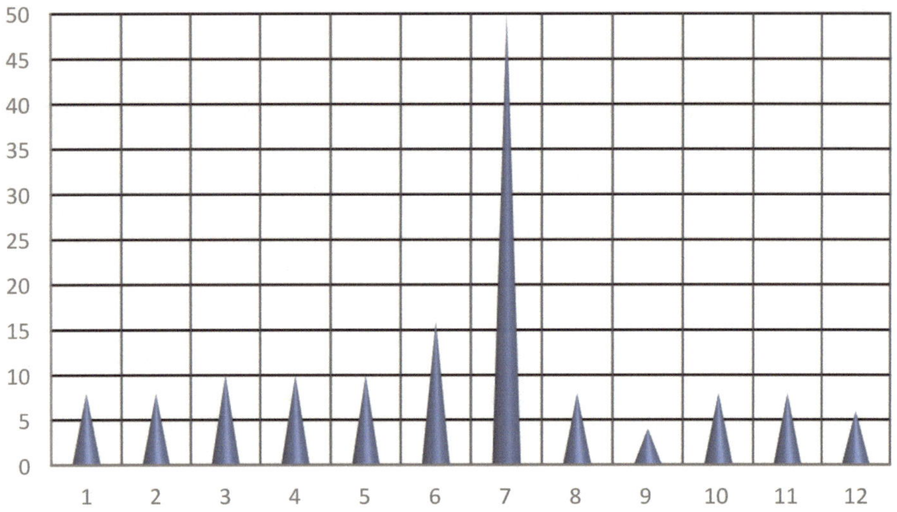

Figure 2.
The upper limits of the range of the maximum energy spectrum of the noise: Along the abscissa axis—The noise source (1—Metallurgical industry, 2—Gas and oil industry, 3—aviation industry, 4—Mining and construction industry, 5—Spacecraft, 6—Turbojet aircraft, 7—Piston aircraft, 8—Helicopters, 9—Motor transport, 10—Railway transport, 11—Cargo river and sea vessels, 12—Hydrofoils and hovercraft) along the ordinate axis—Noise frequency, Hz.

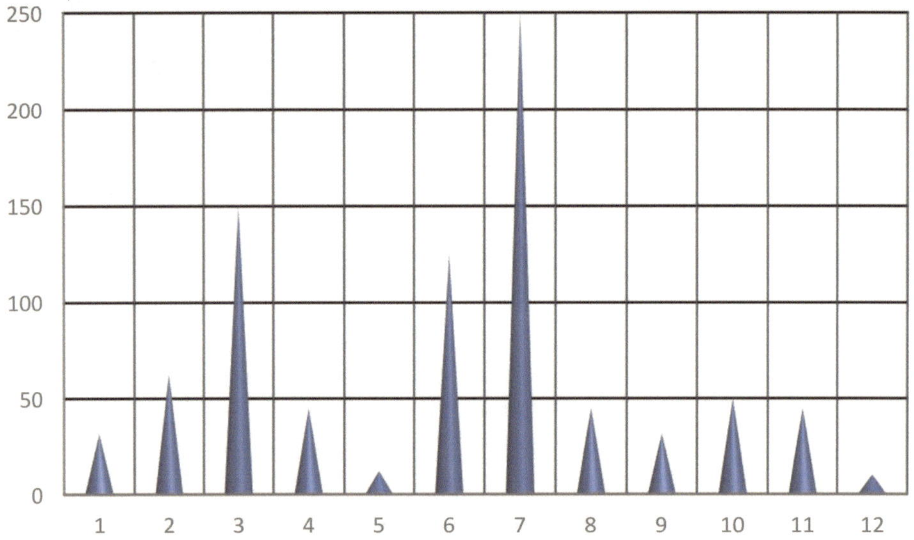

Figure 3.
The lower limits of the range of the maximum energy spectrum of the noise: Along the abscissa axis—The noise source (legend—See ***Figure*** *2).*

ranges. The highest noise levels were noted at the workplaces of motor test stations (SPL 120–132 at frequencies of 50–150 Hz) [8, 16–18].

The main sources of noise in the mining and construction industries are compressors, diesel and ventilation units, vibration platforms, etc. At workplaces, SPL at frequencies of 10–45 Hz is 98–123 dB. In the noise spectrum of vibratory platforms with a high load capacity, SPL in the octave bands of 2–16 Hz is about 100 dB, cranes – 8–16 Hz (79–94 dB), hammers and presses – 8–31.5 Hz (108–114 dB) [7].

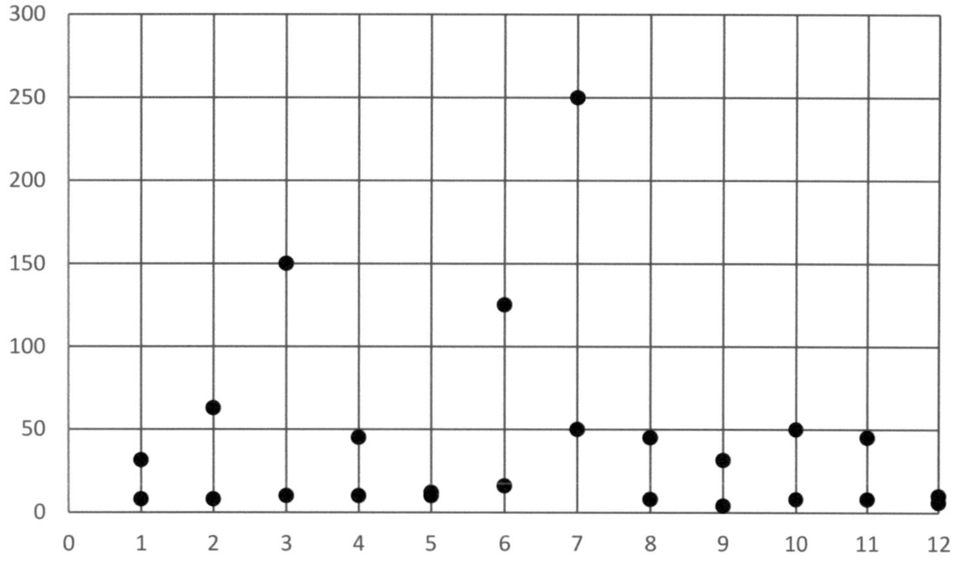

Figure 4.
The upper and lower limits of the range of the maximum energy spectrum of the noise: Along the abscissa axis—The noise source (legend—See ***Figure 2****).*

Jet engines of rockets and airplanes are powerful sources of low-frequency noise and infrasound. When launching rockets of some types, the highest SPL (150 dB or more) are determined at frequencies of 10–12.5 Hz. During takeoff of turbojet aircraft of the TU-154 type with a total noise in the cabins of about 100 dBA, the infrasound levels are 80 dB at a frequency of 4 Hz and 90 dB at a frequency of 20 Hz. In helicopter cockpits, the highest SPLs are 110–120 dB at a frequency of 28 Hz, which corresponds to the rotational speed of the propeller blades. When servicing aircraft with running main and auxiliary engines and ground equipment, aviation specialists at their workplaces are exposed to noise with an SPL of 100–120 dB in octave bands from 2 to 31.5 Hz [8]. Ground vehicles are also significant sources of low-frequency noise and infrasound. Thus, acoustic oscillations with a SPL of 93–110 dB in the range of 8–31.5 Hz are especially characteristic of the cabs of heavy trucks and busses. With fully open windows, an increase in SPL is noted at frequencies of 2–6 Hz. The speed of traffic has a great influence on acoustic performance [19, 20].

In railway transport, the sources of low-frequency noise and infrasound are the power plants of diesel locomotives and electric locomotives, compressor and ventilation units, and aerodynamic flows at high speeds. Railway workers in their workplaces are exposed to noise with SPLs of 92–115 dB at frequencies of 8–50 Hz. Locomotive crews are in the most unfavorable conditions, at the workplaces of which infrasound reaches SPL from 100 to 115 dB. The presence of open windows during the movement of rolling stock leads to an increase in SPL and a shift in the spectrum to the region of low-frequency noise and infrasound, especially at high speeds [7, 21, 22].

Sources of low-frequency noise and infrasound on sea and river vessels are power plants, diesel generators, propellers, ship ventilation and air conditioning systems, etc. Metal hull structures have high sound conductivity, which contributes to the spread of noise throughout the vessel. At workplaces, the seafarers are exposed to noise with an SPL of 100–130 dB at frequencies of 8–45 Hz. The highest noise levels (up to 100 dBA) are observed in the power compartments of ships, which, as a rule, are 30–40 dB higher than in other habitable spaces. During the operation of hydrofoils and hovercraft, SPLs in the region of 6–10 Hz reach 100–130 dB [7].

2.3 Means of measurement and hygienic regulation of industrial and transport noise

Acoustic measurements are mainly carried out for research purposes, designed to reduce the level of influencing noise, for hygienic monitoring, to comply with the requirements of sanitary standards, as well as for predictive risk assessment of the development of diseases associated with acoustic exposure [1, 7, 23–25].

For hygienic standardization according to the temporal characteristics of noise, gradation is introduced into three categories [1]:

- constant noise, the sound level of which does not change by more than 5 dBA during an 8-hour working day or during the measurement in the averaging mode of the sound level meter S (slow);

- intermittent noise, the sound level of which changes by more than 5 dBA during an 8-hour working day, a work shift or during a measurement when measured with the averaging time constant of the sound level meter S (slowly);

- impulse noise, consisting of the infrasound of one or more sound events, each with a duration of less than 1 s, while the sound levels $L_{p,AImax}$ and $L_{p,ASmax}$, measured respectively with time corrections I (impulse) and S (slow), differ by at least 7 dB.

The current standards for hygienic regulation of noise at the workplace of personnel use the following indicators:

sound pressure level, L_p, dB is 10 decimal logarithms of the ratio of the square of the sound pressure to the square of the reference sound pressure equal to 20 μPa;
the equivalent sound pressure level, $L_{p,eqT}$, dB is 10 decimal logarithms of the ratio of the square of the sound pressure to the square of the reference sound pressure over a given time interval;
A-weighted sound level (A sound level) in dBA is 10 decimal logarithms of the ratio of the squared rms sound pressure, measured using standardized frequency weighting A, to the square of the reference sound pressure. To determine the nature of the noise, sound levels A are measured with time corrections S (slow, $\tau = 1$ s), F (fast, $\tau = 125$ ms), I (impulse, $\tau = 40$ ms);

equivalent sound level with frequency correction A (equivalent sound level A), $L_{p,Aeq,T}$, dBA—10 decimal logarithms of the ratio of the square of the root-mean-square sound level A to the square of the reference sound pressure at a given time interval, which is calculated by the formula:

$$L_{p,Aeq,T} = 10lg\left[\frac{\int_{t_1}^{t_2} P_A^2(t)dt}{TP_0^2}\right], \tag{1}$$

equivalent sound level A for a work shift—$L_{p,Aeq,8h}$, dBA, equivalent sound level A, measured or calculated for 8 hours of a work shift, taking into account corrections for impulse and tonal noise, which is calculated by the formula:

$$L_{p,Aeq,8h} = 10lg\left(\frac{1}{T_0}\sum_i T_i 10^{0,1\left(L_{p,Aeq,Ti}+K_i\right)}\right), \quad (2)$$

where T_0 is the standard duration of the work shift (8 hours, if the duration of the work shift is different from 8 hours, T_0 is taken equal to the actual duration of the work shift with a total work duration of 40 hours per week); T_i is the duration of the i-th noise exposure interval, h; $L_{p,Aeq,Ti}$ is the equivalent sound level or sound pressure measured at the i-th noise exposure interval, dBA; K_i is a correction for the nature of the noise, equal to 5 dB in the case of tonal and (or) impulse noise (applied when $L_{p,Aeq,Ti} > 75$ dBA, in all other cases $K = 0$ dB is assumed);

the maximum sound level A, $L_{p,Amax}$, dBA is the highest sound level measured over a given time interval with standard time correction;

the time correction function is a standard exponential function of time for the square of the instantaneous sound pressure in a time averaging operation (international standard). Sound level meters use standard time corrections S (slow, $f = 1$ s), F (fast, $f = 125$ ms), I (impulse, $f = 40$ ms), they are also called averaging time constants;

Peak C-weighted sound level (C sound level), $L_{p,Cpeak}$, dBS, is 10 decimal logarithms of the ratio of the square of the peak sound pressure, measured using standardized frequency equalization, to the square of the reference sound pressure.

It should be noted that in addition to the equivalent sound level, when describing and characterizing short-term sounds or noises, in the international practice of acoustic measurements (not standardized in Russia), the exposure parameter of a separate noise phenomenon (event) is often used [15]. The equivalent sound exposure level is the level that, when fixed for a time interval of 1 s, produces sound energy in dBA that is identical to the energy of an actual transient sound or noise. This level, in dBA, is calculated using the formula:

$$L_{AE} = 10\left(\frac{1}{t_0}\int_{t_1}^{t_2}\frac{P_A^2(t)}{P_0^2}dt\right), \quad (3)$$

where t_0 is the standard duration equal to 1 s.

For the theory and practice of hygienic regulation of the noise factor, acoustic parameters characterizing the average sound level over a long period of time are of undoubted significance. A long time interval consists of infrasound series of basic intervals. A day or some other time period characterizing a certain technological cycle of human activity is used as a base interval. To quantify the noise factor, the average sound level over a long time interval is determined—the average value of the equivalent sound level over a long time interval for a series of basic time intervals enclosed within a long time interval:

$$L_{AeqLT} = 10lg\left(\frac{1}{N}\sum_{i=1}^{N}10^{0,1\left(L_{Aeq,T}\right)_i}\right), \quad (4)$$

where N is the number of basic time intervals, $(L_{Aeq,T})_i$ is the equivalent SPL in the i-th basic time interval.

Another parameter often used for the purposes of hygienic regulation and health risk assessment is the noise dose—the total energy accumulated during exposure. The noise dose is proportional to the equivalent (in terms of energy) sound pressure recorded on the frequency correction scale "A" and the action time, measured in Pa^2h or Pa^2s. Dose—acoustic energy during the duration of the noise, determined by the formula:

$$Dose = \int_0^T P_A^2(t)dt, \tag{5}$$

where P_A is the instantaneous value of sound pressure on the A scale of the sound level meter, Pa; T is the measurement time, h.

Sometimes it is more convenient to use the relative value of the noise dose (Dsh) in fractions of the permissible value (dimensionless value):

$$D_{sh} = \frac{Dose}{Dose_{dop}}; Dose_{dop} = P_{dopA}^2 T_{dop}, \tag{6}$$

where $P_{dop\ A}$, Pa—the permissible value of sound pressure on the A scale of the sound level meter, corresponding to the maximum permissible noise level of 80 dBA, T_{dop}—the established duration of the working day.

From a physical point of view, the noise dose reflects the amount of energy transferred and thus can serve as a measure of exposure. Therefore, such a hygienic aspect of its application is important, such as the rigor of calculating the noise estimate in comparison with the current system of normalizing its level.

Dose estimation has an undeniable advantage: it takes into account the transferred energy during the noise action, which allows one to estimate the noise load and correlate it with the biological effects caused. In this regard, its use helps to identify qualitative-quantitative relationships of the fundamental dose-effect ratio.

The noise dose (Dsh) is related to the equivalent L_{eq}, the maximum permissible sound levels L_{dop} and the corresponding sound pressure values P_{ekv}, P_{dop} by the relation:

$$D_{sh} = 10^{\frac{(L\,eq - L\,dop)}{10}} = 10^{\frac{\left(L\,izm\ A + 10\,\lg\left(\frac{Tc}{To}\right) - L\,dop\right)}{10}} = \left(P_{eq}/P_{dop}\right)^2 = Dose/Dose_{dop}. \tag{7}$$

The transition from sound levels measured in dB to the root mean square value of the sound pressure and its square was carried out according to the following formulas:

$$P_{Aizm} = 2 \times 10^{[L/20]-5}, \tag{8}$$

$$P_{A\ izm}^2 = 4 \times 10^{[\,L/10]-10} \tag{9}$$

Based on the physical concept of noise dose, the expression for the total dose per shift has the form:

$$D_{sh} = \sum_i D_{shi}, \tag{10}$$

where D_{shi}—partial doses of noise; i is an index denoting stages of work or basic intervals.

The dose approach to determining the acoustic load is also convenient to use to calculate the integral estimate of the entire impact. It is necessary to additionally determine the actual and relative noise dose for working hours, rest periods and sleep, i.e. for all periods of life. The relative noise dose, separately for working hours, rest periods and sleep is determined by the formula:

$$ODsh = D_{\text{otn}} = \text{Dsh}_{\text{fact}}/\text{Dsh}_{\text{dop}} = \frac{\sum_{i=1}^{n} P_i^2 t_i}{Dsh_{dop}}, \tag{11}$$

where P_i—sound pressure in Pa; t_i is the duration of a given noise level (working time, rest, sleep), D_{rel} is the relative noise dose, Dsh_{fact} is the actual noise dose, Dsh_{dop}—permissible noise dose.

The total relative average daily noise dose is determined as the sum of the ratios of the real noise load for three periods of life (work, rest and sleep) to the permissible one according to the formula:

$$ODsh = \frac{Dsh_{job}}{Dsh_{dop.job}} + \frac{Dsh_{relaxation}}{Dsh_{dop.relaxation}} + \frac{Dsh_{dream}}{Dsh_{dop.dream}}. \tag{12}$$

It can be noted that for research purposes, to build the "dose-effect" dependence, the noise dose measured over the entire frequency range on the "Linear" scale is used.

The normalized indicators of noise at workplaces are:

- equivalent sound level A for a work shift, the normative equivalent sound level at workplaces is 80 dBA;

- maximum sound levels A, measured with time corrections S and I, maximum sound levels A, measured with time corrections S and I, should not exceed 110 dBA and 125 dBA, respectively.

- peak sound level C should not exceed 137 dBC.

Exceeding any normalized parameter is considered to be exceeding the maximum permissible levels.

For certain sectors (sub-sectors) of the economy, an equivalent noise level at workplaces from 80 to 85 dBA is allowed, subject to the confirmation of an acceptable risk to the health of workers based on the results of calculations of the assessment of the occupational risk to the health of workers, as well as the implementation of a set of measures aimed at minimizing the risks to the health of workers.

If the noise level in the workplace exceeds 80 dBA, the employer must carry out an assessment of the health risk of workers and confirm the acceptable health risk.

Work in conditions of exposure to an equivalent noise level above 85 dBA is not allowed.

The impact of noise on a person depends on the intensity, frequency composition and duration of its action, as well as on the location of the person and the nature of the work. Noise with a level of 30–40 dBA at night can cause anxiety, insomnia; at 50–60 dBA, if a person is engaged in mental work, a load is created on the nervous system, and a harmful psychological effect is observed. Sound levels up to 70 dBA already cause certain physiological reactions and can lead to changes in the body. Noise, the US of which reaches 80–90 dBA, affects hearing, causing its deterioration, and high sound levels cause the development of a specific occupational disease—industrial sensorineural hearing loss.

In the current standards for hygienic regulation of infrasound at workplaces, the following indicators are used:

Total infrasound sound pressure level (total infrasound level): sound pressure level in the frequency range 1.4–22 Hz, can be directly measured with an appropriate band pass filter or obtained by energy summing the sound pressure levels in the octave frequency bands 2, 4, 8, 16 Hz.

Equivalent sound pressure level, $L_{p,eq,T}$, dB—10 decimal logarithms of the ratio of the square of the sound pressure to the square of the reference sound pressure over a given time interval. Equivalent sound pressure levels for a shift in octave frequency bands are determined by the formula:

$$L_{p,1/1,eq,8h} = 10lg\left(\frac{1}{T_0}\sum_i T_i 10^{0,1L_{p,1/1,eq,T_i}}\right), \quad (13)$$

where T_O is the standard duration of the work shift (8 hours). If the duration of the work shift is different from 8 hours, T_O is taken equal to the actual duration of the work shift with a total work duration of 40 hours per week. T_i is the duration of the i-th interval of infrasound exposure, h; $L_{p,1/1,eq,Ti}$ is the equivalent sound pressure level measured in the i-th interval, dB in the octave band.

The equivalent total infrasound level for a work shift is determined by the formula:

$$L_{p,ZI,eq,8h} = 10lg\left(\sum_{i=1}^{n} T_i 10^{0,1L_{p,ZI,eq,T_i}}\right), \quad (14)$$

where T_O is the standard duration of the work shift (8 hours); T_i is the duration of the i-th interval of infrasound exposure, h; $L_{p,ZI,eq,8h}$—replaceable equivalent total infrasound level; $L_{p,ZI,eq,Ti}$ is the equivalent total infrasound level measured at the i-th interval of its impact.

The maximum sound pressure level $L_{p,max}$, dB is the highest value of the sound pressure level measured over a given time interval with standard time correction (time constant).

The normalized parameters of infrasound are:

- equivalent sound pressure levels for a working shift in octave frequency bands 2, 4, 8, 16 Hz—$L_{p,1/1,eq,8h}$, dB;
- equivalent total infrasound level per shift—$L_{p,ZI,eq,8h}$, dB—should not exceed 100 dB;
- maximum total infrasound level, measured with time correction S (slow)—should not exceed 120 dB.

To obtain an approximate assessment of the severity of infrasound, you can use the overall sound level measured on the "Lin" scale, and the express indicator—the difference between the sound levels measured on the "Lin" and "A" scales. The greater this difference, the more significant the contribution of low-frequency and infrasonic components in the spectrum of the studied noise. At values of the indicator from 6 to 10 dB, it is considered that there are signs of the presence of infrasound, at 11–20 dB, infrasound is moderately pronounced; 21–30 dB—expressed; more than 30 dB—significant infrasound.

The hygienic standard refers to the quantitative and qualitative values of indicators established by research that characterize the safety of environmental factors for human health. When regulating noise and infrasound, a multi-level approach was used, depending on the nature of human activity. The existing sanitary rules establish the maximum permissible levels at workplaces, permissible levels in residential and public premises and in residential areas. The determination of the maximum permissible levels at the workplace should be made taking into account the intensity and severity of labor activity [7, 8].

The regulation of normative noise values is based on preventing the development of irreversible hearing impairment. At the same time, for infrasound, the approach of its general impact on a person is used, taking into account the reaction of the hearing organ. This has led to the fact that at present there are significant differences in the values of standard SPL in octave bands with geometric mean frequencies of 16 Hz (85 dB) in the infrasonic range and 31.5 Hz (107 dB) and 63 Hz (95 dB) in the low-frequency audio range. It has been established that low-frequency noise can have a harmful effect not only on hearing, but also on other human systems, and its biological effect is similar to the effect of infrasound on a person. Therefore, studies are needed to clarify the nature of the frequency dependence of biological effects on the conditional boundary between the infrasonic and sonic ranges [24, 26–28].

The above examples show the need for further research to justify the equally effective values of the weighting coefficients under the influence of low-frequency noise and infrasound [29, 30].

2.4 Working conditions

Working conditions are a combination of factors of the labor process (severity and tension) and the working environment (physical, chemical, and biological factors) in which human activities are carried out. The classification of working conditions is based on the principle of grading the deviation of the parameters of these factors from the current hygienic standards [1, 7, 8, 24, 31].

When evaluating working conditions associated with the action of several harmful factors, the summation effect is taken into account depending on the number of factors and the severity of their harmfulness.

In addition, if there is both noise and infrasound at the workplace, working conditions should be rated one step higher. The legitimacy of this approach is due to the fact that these two factors can have a harmful effect on the same critical organs and systems ("targets"), which leads to the summation and potentiation of their adverse effects.

2.5 Medical aspects

For a long time it was believed that infrasound lies beyond the limits of auditory perception. It has now been established that they are perceived not as pure tones, but as a combination of auditory and tactile sensations, which is manifested by a feeling of pulsation in the tympanic membrane and middle ear. The hearing thresholds for infrasound have been established: for 100 Hz they are about 40 dB, and for 1 Hz −140 dB [1, 31].

Long-term action of low-frequency noise and infrasound leads to an increase in the threshold of hearing, mainly in the low and medium frequency ranges. Considering that the maximum of speech frequencies is in these areas, these disorders are prognostically unfavorable in social terms, which is especially pronounced in socio-professional groups of the population exposed to aircraft noise [7, 32–37].

A questionnaire survey of workers exposed to low-frequency noise and infrasound in production and transport for a long time revealed a complex of unpleasant subjective sensations in the majority [1, 7, 8, 24, 31]. Complaints, depending on their genesis, can be conditionally divided into the following groups:

cochlear: a feeling of congestion, pressure, pulsation and pain in the ears, hearing impairment;

vestibular: dizziness, imbalance; mechanical: sensation of vibration of the chest and abdominal wall, soft palate, internal organs, cough, shortness of breath, blurred vision; psychological: anxiety, unreasonable feeling of fear, decreased mood, apathy, problems with concentration and memory;

neurovegetative: fatigue, general malaise, drowsiness, irritability, sleep disturbances, headache, dizziness, loss of appetite, tachycardia, fluctuations in blood pressure.

The variety of complaints indicates the involvement of many organs and systems in the formation of the subjective perception of low-frequency noise and infrasound [1, 7, 24, 31].

The presence of harmful factors, having an adverse effect on the body of workers, leads to an increase in the level of chronic and general, production-related and occupational morbidity [7].

The impact of noise with a low-frequency and infrasound component is accompanied by an increase in the general morbidity and diseases characteristic of the action of noise and infrasound. This indicates the summation of adverse effects in the combined influence of noise and infrasound. In the structure of morbidity, diseases of the organs of hearing, respiration, blood circulation, digestion, skin and subcutaneous tissue, and the nervous system predominate, and the leading place among them is occupied by sensorineural hearing loss and arterial hypertension [1, 7, 8, 24, 31].

The diseases identified in specialists are related to working conditions based on quantitative assessments of occupational risk, which makes it possible to classify diseases of the organ of hearing as occupational diseases, and diseases of the respiratory organs, eyes, digestion, nervous system, circulatory organs and skin as occupationally caused diseases [31].

At the end of the last century, an understanding was formed that exposure to harmful factors (including noise and infrasound) can lead to the development of occupational diseases. Russia has created a system of medical support for people who are exposed to harmful factors at work. The basis of the complex of medical measures is the conduct of preliminary examinations when applying for a job and periodic medical examinations. In particular, it provides for the passage of persons exposed to noise and infrasound at work, periodic medical examinations (at least once every 2 years) with the obligatory participation of an otolaryngologist and a neurologist, laboratory tests of auditory and vestibular analyzers [1, 7, 8, 31]. According to the indications, an examination and examination of the connection of the disease with the profession is carried out in the conditions of a specialized authorized medical organization.

One of the promising directions for ensuring the acoustic safety of personnel is the implementation of personalized acoustic monitoring technologies based on the use of real-time acoustic hazard indicators [38, 39]. Realized with the help of personalized acoustic monitoring and medical control system, the accumulation of factual information about the influence of acoustic factors on a person, an adequate and reliable quantitative description of the patterns of changes in health and performance open up new opportunities for testing, correcting and justifying management decisions aimed at maintaining health and prolonging professional longevity. Workers exposed to industrial and traffic noise [38, 39].

An important aspect of industrial and transport noise monitoring is noise mapping [40–42]. The construction of noise maps, especially those that provide objective monitoring of noise pollution in real conditions with a possible determination of the potential effectiveness of the implementation of a set of measures aimed at reducing noise pollution, is an effective tool to support the adoption of appropriate management decisions [43, 44].

2.6 Methods and means of protection against industrial and transport noise

When choosing means and methods of protection against low-frequency noise and infrasound, it must be borne in mind that: there are actually no specialized means of protection against infrasound; in industrial conditions, infrasound is often combined with intense noise; most personal protective equipment designed to protect the organ of hearing is ineffective at frequencies below 500 Hz (sound attenuation does not exceed 10 dB).

The variety of applied methods and means of protection against industrial and transport noise necessitates the development of special information-measuring systems for their qualimetry [45, 46]. Expanding the spectrum of industrial and traffic noise to the low-frequency and infrasonic region, as well as taking into account the parameters of impulse noise, requires a new solution of acoustic problems that have not had a theoretical solution so far. So, there are still no exact methods for solving the propagation of a sound wave over remote distances, there are only probabilistic-statistical approaches applicable exclusively to a specific area and relief where acoustic measurements were made [47].

For timely decision-making on protection from industrial and traffic noise, it is necessary to have objective and comprehensive information about the characteristics of the noise environment [48, 49]. Therefore, the creation of an effective system of dosimetric control, that is, monitoring of noise levels, is one of the urgent tasks of measuring systems for qualimetry of means and methods of noise protection.

The functional tasks of the complex of information-measuring systems for qualimetry of means and methods of protection against industrial and transport noise are: (1) reducing the level of personnel noise dose to regulated limits based on a set of design, technical and health measures; and (2) creation of an effective system of dosimetric control, which makes it possible to quickly register an increase in the level of acoustic exposure to a person who is exposed to noise.

When choosing personal protective equipment, you should be guided by the following.

1. In the presence of noise, the spectrum of which is dominated by medium and high frequencies, and the SPL of low-frequency noise and infrasound do not exceed the maximum permissible levels, it is necessary to use anti-noise

(headphones, earbuds and a helmet) designed to protect the hearing organ. When choosing anti-noise, it should be taken into account that at the noise level:

up to 100 dBA—you need to use headphones or earbuds;

100–110 dBA—a combination of headphones with earbuds;

110–125 dBA—anti-noise helmets, vests, suits.

2. When exposed to infrasound with levels exceeding the maximum permissible levels, and intense noise, it is necessary to protect not only the hearing organ, but also the central and autonomic nervous systems, the cardiovascular system, and the respiratory system. This is achieved by special noise protection equipment—anti-noise helmet, vest and suit [8, 50].

Special noise protection equipment is a new class of technical personal protective equipment designed to protect a person from the extracochlear effects of infrasound and low frequencies of the sound range. Industrial samples of headphones and experimental samples of anti-noise helmets and vests, which reduce the level of acoustic energy in the low-frequency range, have been developed [8].

An important role in ensuring protection against low-frequency noise and infrasound in the workplace belongs to measures to optimize the conditions of professional activity—the use of collective protective equipment, reducing the length of stay in the noise zone, alternating periods of work and rest, etc. It is necessary to use the alternation of work periods for maintenance of production equipment ("active period of acoustic load"), with periods not related to the maintenance of noise sources ("passive period of acoustic load"). In the latter case, it is important to create comfortable acoustic conditions and carry out rehabilitation measures [8].

Thus, the use of protective equipment is necessary for the effective prevention of occupational morbidity, and hence the reduction of economic losses in production.

3. Conclusion

Industrial facilities and most modes of transport are sources of high-intensity noise, the spectrum of which is dominated by low-frequency infrasonic frequencies. The close physical nature of these ranges contributes to the propagation of such noise with low attenuation, and they have good penetrating power, so most noise protection devices are ineffective. High (more than 100 dBA) noise levels at workplaces of industrial facilities and transport require measurements in the infrasound range as well.

Studies of low-frequency acoustic oscillations as a factor in the production environment have not been completed. Their "targets" are the central nervous and autonomic nervous systems, auditory and vestibular analyzers, respiratory organs, etc. With prolonged exposure, they contribute to the development of occupational diseases. The simultaneous action of low-frequency noise and infrasound (this situation is typical for industrial conditions and vehicles) leads to an aggravation of infrasound pathology, which requires more careful medical monitoring of people working in such conditions, and improvement of means and methods of protection against industrial and transport noise.

Acknowledgements

This work was supported by a grant from the President of the Russian Federation for the state support of leading scientific schools of the Russian Federation (grant NSh-122.2022.1.6).

Conflict of interest

The authors declare no conflict of interest.

Author details

Sergey Dragan and Aleksey Bogomolov*
State Research Center, Burnasyan Federal Medical Biophysical Center of Federal Medical Biological Agency, Moscow, Russia

*Address all correspondence to: a.v.bogomolov@gmail.com

References

[1] Uiba VV, Kotenko KV, Stepanov VS. Biological Effect of Infrasound. Moscow: FMBA; 2012. p. 384

[2] Prokopenko LV, Kurierov NN, Lagutina AV, Pochtareva ES. Substantiation of risk-oriented hygienic criteria and classification of working conditions by noise taking into account the group attributive (excessive) risk of hearing loss. Occupational Medicine and Industrial Ecology. 2022;**62**(3):185-192. DOI: 10.31089/1026-9428-2022-62-3-185-192

[3] Grubesa S, Suhanek M. Traffic noise. In: Siano D, González AE, editors. Noise and Environment. London: IntechOpen; 2020. DOI: 10.5772/intechopen.92892

[4] Bukhtiyarov IV, Denisov EI, Couriers NN, Prokopenko LV, Bulgakova MV, Khakhileva OO. Improving the criteria for noise hearing loss and occupational risk assessment. Occupational Medicine and Industrial Ecology. 2018;**4**:1-9

[5] Benocci R, Bisceglie A, Angelini F, Zambon G. Influence of traffic noise from local and surrounding areas on high-rise buildings. Applied Acoustics. 2020;**166**:107362. DOI: 10.1016/j.apacoust.2020.107362

[6] Przysucha B, Szeląg A, Pawlik P. Probability distributions of one-day noise indicators in the process of the type A uncertainty evaluation of long-term noise indicators. Applied Acoustics. 2020;**161**:107158. DOI: 10.1016/j.apacoust.2019.107158

[7] Izmerov NF, Suvorov GA, Kuralesin NA, Ovakimov VG. Infrasound as a Risk Factor for Human Health (Hygienic, Biomedical and Pathogenetic Mechanisms). Voronezh: Istoki; 1998. p. 276

[8] Soldatov SK, Zinkin VN, Bogomolov AV, Dragan SP, Kukushkin YA. Fundamental and Applied Aspects of Aviation Medical Acoustics. Moscow: Fizmatlit; 2019. p. 214

[9] Brown AL, van Kamp I. WHO environmental noise guidelines for the European region: A systematic review of transport noise interventions and their impacts on health. International Journal of Environmental Research and Public Health. 2017;**14**(8):873. DOI: 10.3390/ijerph14080873

[10] Bukhtiyarov IV, Kurierov NN, Lagutina AV, Prokopenko LV, Zibarev EV. Aircraft noise in residential buildings, problems of measurement and evaluation. Hygiene and Sanitation. 2020;**99**(10):1042-1048. DOI: 10.47470/0016-9900-2020-99-10-1042-1048

[11] Dragan SP, Bogomolov AV. Method for assessing human acoustic safety. Siberian Journal of Life Sciences and Agriculture. 2021;**13**(1):259-278. DOI: 10.12731/2658-6649-2021-13-1-259-278

[12] Alam P, Ahmad K, Afsar SS, Akhtar N. Noise monitoring, mapping, and modelling studies-a review. Journal of Ecological Engineering. 2020;**21**(4): 82-93. DOI: 10.12911/22998993/119804

[13] Mccomas S, Arrowsmith S, Hayward C, Stump B, Taylor MHM. Quantifying low-frequency acoustic fields in urban environments. Geophysical Journal International. 2022; **229**(2):1152-1174. DOI: 10.1093/gji/ggab525

[14] Dragan SP, Bogomolov AV, Zinkin VN. Methodical support of monitoring the acoustic safety of flight personnel. AIP Conference Proceedings. 2019;**2140**:020019. DOI: 10.1063/1.5121944

[15] Shetty D, Prabhu N. Power spectral density based identification of low frequency oscillations in multimachine power system. Journal of Operation and Automation in Power Engineering. 2023;**11**(3):173-181. DOI: 10.22098/joape.2023.10072.1710

[16] Dragan SP, Drozdov SV, Soldatov SK, Bogomolov AV, Shishov AA, Zinkin VN, et al. Substantiation of methodological approaches to determining the boundaries of the sanitary protection zones of the near-aerodrome territory according to the level of noise impact. Aerospace and Environmental Medicine. 2021;**55**(5):85-93. DOI: 10.21687/0233-528X-2021-55-5-85-93

[17] Basner M, McGuire S. WHO environmental noise guidelines for the european region: A systematic review on environmental noise and effects on sleep. International Journal of Environmental Research and Public Health. 2018;**15**(3):519. DOI: 10.3390/ijerph15030519

[18] Truong TP, Nguyen NT. Towards an iot-based acoustic monitoring system in smart cities. ARPN Journal of Engineering and Applied Sciences. 2022;**17**(6):640-644

[19] Srivastava S, Mishra AK. Study of noise pollution in National Highway-19, Mathura City. Indian Journal of Environmental Protection. 2022;**42**(6):722-729

[20] Tang JH, Lin BC, Hwang JS, Chen LJ, Wu BS, Jian HL, et al. Dynamic modeling for noise mapping in urban areas. Environmental Impact Assessment Review. 2022;**97**:106864. DOI: 10.1016/j.eiar.2022.106864

[21] Pankova VB, Vilk MF, Zibarev EV, Fedina IN. On the question of accounting for new factors in the pathogenesis of occupational hearing loss (by the example of transport workers). Occupational Medicine and Industrial Ecology. 2022;**62**(8):488-500. DOI: 10.31089/1026-9428-2022-62-8-488-500

[22] Zambon G, Benocci R, Bisceglie A, Roman HE, Bellucci P. The LIFE DYNAMAP project: Towards a procedure for dynamic noise mapping in urban areas. Applied Acoustics. 2017;**124**:52-60. DOI: 10.1016/j.apacoust.2016.10.022

[23] Rajamanickam R, Nagan S. Status of ambient noise level in Chennai City and strategy for mitigation. Indian Journal of Environmental Protection. 2022;**42**(4):415-423

[24] Karpova NI, Malyshev EN. Low-Frequency Acoustic Vibrations. Moscow: Medicina; 1981. p. 192

[25] Bogomolov AV, Dragan SP. A new approach to the study of impedance characteristics of tympanic membrane. Doklady Biochemistry and Biophysics. 2015;**464**(1):269-271. DOI: 10.1134/S1607672915050014

[26] Bustaffa E, Curzio O, Donzelli G, Bianchi F, Minichilli F. Risk associations between vehicular traffic noise exposure and cardiovascular diseases: A residential retrospective cohort study. International Journal of Environmental Research and Public Health. 2022;**19**(16):10034. DOI: 10.3390/ijerph191610034

[27] Mann S, Singh G. Traffic noise monitoring and modelling. Environmental Science and Pollution Research. 2022;**29**(37):55568-55579. DOI: 10.1007/s11356-022-21395-4

[28] Rojek M, Wojciechowska W, Januszewicz A, Skalski P, Rajzer M. The relation of nocturnal exposure to aircraft noise and aircraft noise-induced insomnia with blood pressure. Polish Archives of Internal Medicine. 2021; **131**(1):33-41. DOI: 10.20452/pamw.15716

[29] Gulliver J, Morley D, Vienneau D, Fabbri F, Bell M, Goodman P, et al. Development of an open-source road traffic noise model for exposure assessment. Environmental Modelling and Software. 2015;**74**:183-193. DOI: 10.1016/j.envsoft.2014.12.022

[30] Bogomolov AV, Gan SP, Zinkin VN, Alekhin MD. Acoustic factor environmental safety monitoring information system. In: Acoustic Factor Environmental Safety Monitoring Information System, XXII International Conference on Soft Computing and Measurements (SCM). St. Petersburg: LETI; 2019. pp. 215-218. DOI: 10.1109/SCM.2019.8903729

[31] Akhmetzyanov IM, Grebenkov SV, Lomov OP. Noise and infrasound. In: Hygienic Aspects. St. Petersburg: Bip; 2002. p. 100

[32] Drozdov SV, Dragan SP, Bogomolov AV, Suleymanov AE. A method to determine aircraft speed from ground acoustic measurements. Moscow University Physics Bulletin. 2020;**75**(5): 459-464

[33] Simons DG, Besnea I, Mohammadloo TH, Melkert JA, Snellen M. Comparative assessment of measured and modelled aircraft noise around Amsterdam airport schiphol. Transportation Research Part D: Transport and Environment. 2022; **105**:103216. DOI: 10.1016/j.trd.2022.103216

[34] Makino K, Shinohara N. Daily fluctuations in aircraft noise exposure around civil airports and military airfields in Japan. In: Proceedings of INTER-NOISE 2021. Washington, D.C., USA: International Congress and Exposition of Noise Control Engineering; 2021. pp. 5154-5160. DOI: 10.3397/IN-2021-2984

[35] Vahl JM, Keppeler JVA, Krahe D, Hoffmann TK, Goldberg-Bockhorn E. Infrasound—Implications for human medicine. HNO. 2022;**70**(12):921-930. DOI: 10.1007/s00106-022-01237-y

[36] Ganić E, Ivošević J, Mirković B. Impact of aircraft noise on communities near Belgrade airport. Promet-Traffic-Traffico. 2021;**33**(3):323-335. DOI: 10.7307/ptt.v33i3.3692

[37] He Y, Li J, Tian Y, Qian J, Yang Y. Optimization and effect comparison of aircraft noise automatic monitoring sites based on comprehensive weight-TOPSIS. Chinese Journal of Environmental Engineering. 2022;**16**(6): 2021-2031. DOI: 10.12030/j.cjee.202202012

[38] Ushakov IB, Bogomolov AV, Dragan SP, Soldatov SK. Methodological foundations of personalized acoustic monitoring. Labor Safety in Industry. 2020;**10**:33-39. DOI: 10.24000/0409-2961-2020-10-33-39

[39] Ushakov IB, Bogomolov AV, Dragan SP, Soldatov SK. Methodological bases of personalized hygienic monitoring. Aerospace and Environmental Medicine. 2017;**51**(6): 53-56. DOI: 10.21687/0233-528X-2017-51-6-53-56

[40] Saucy A, Schäffer B, Tangermann L, Wunderli JM, Röösli M. Individual aircraft noise exposure assessment for a case-crossover study in Switzerland. International Journal of Environmental Research and Public Health. 2020;**17**(9): 3011. DOI: 10.3390/ijerph17093011

[41] Ushakov I, Bogomolov A, Dragan S, Soldatov S. Technology for predictive monitoring of the performance of cyber-physical system operators under noise conditions. In: Kravets AG, Bolshakov AA, Shcherbakov MV, editors. Society 5.0: Cyberspace for Advanced Human-Centered Society. Studies in Systems, Decision and Control. Vol. 333. Cham: Springer; 2021. pp. 269-280. DOI: 10.1007/978-3-030-63563-3_21

[42] Aumond P, Can A, Mallet V, De Coensel B, Ribeiro C, Botteldooren D, et al. Kriging-based spatial interpolation from measurements for sound level mapping in urban areas. Journal of the Acoustical Society of America. 2018; **143**(5):2847-2857. DOI: 10.1121/1.5034799

[43] Kumar D. FL-NoiseMap: A federated learning-based privacy-preserving urban noise-pollution measurement system. Noise Mapping. 2022;**9**(1):128-145. DOI: 10.1515/noise-2022-0153

[44] Benocci R, Roman HE, Zambon G. Optimized sensors network and dynamical maps for monitoring traffic noise in a large urban zone. Applied Sciences. 2021;**11**(18):8363. DOI: 10.3390/app11188363

[45] Bogomolov AV, Dragan SP, Olenina IV. A hardware-software complex for ipsilateral measurements of the acoustic reflex threshold. Instruments and Experimental Techniques. 2021;**4**(64):586-595. DOI: 10.1134/S0020441221040023

[46] Dragan SP, Bogomolov AV, Kotlyar-Shapirov AD, Kondrat'eva EA. Experimental study of displays in contralateral acoustic reflex auditory stimulation. Doklady. Biochemistry and Biophysics. 2016;**468**(1):224-228. DOI: 10.1134/S1607672916030182

[47] Suhanek M, Djurek I, Grubesa S, Petosic A. An advanced approach to calculation of overall soundscape annoyance rating. Forum Acusticum. 2020;**5**(1):1477-1481. DOI: 10.48465/fa.2020.0031

[48] Bogomolov AV, Dragan SP, Erofeev GG. Mathematical model of sound absorption by lungs with acoustic stimulation of the respiratory system. Doklady Biochemistry and Biophysics. 2019;**487**(1):247-250. DOI: 10.31857/S0869-5652487197-101

[49] Serebryakov PV, Saarkoppel LM, Fedina IN, Pankova VB, Bomstein NG. Premorbid effects of impact of the noise-vibration factor. Therapy. 2022;**1**:79-80. DOI: 10.18565/therapy.2022.7suppl.79-80

[50] Dragan SP, Bogomolov AV, Kondrat'eva EA. Mathematical model of polyharmonic signal processing to study the state of the tympanic membrane. In: International Conference on Engineering and Telecommunication (EnT). Moscow: MIPT; 2016. pp. 21-25. DOI: 10.1109/EnT.2016.013

Chapter 4

Evaluation of Industrial Noise Reduction Achieved with a Green Barrier: Case Study

Martha Cobo Dorado, Gissell Rodríguez Milan and Alice Elizabeth González

Abstract

In this chapter, a case study is presented on the evaluation of acoustic performance of a tree barrier. It is a eucalyptus barrier that was planted as a visual barrier to block an industrial plant. First, the depletion law of sound pressure levels (SPL) of the source was analyzed; a linear divergence was found. A calculation scheme similar to that of ISO 9613-2 was applied. When comparing the SPL measured at a specific receiver with the results of propagating the SPL from the source without considering the existence of the barrier, an extra attenuation of 12 dB appeared, reinforcing the idea that the plantation behaves as an acoustic barrier. Four different calculations were used to obtain its insertion loss (IL), including general equations and expressions developed for green barriers. The best fit was obtained using equations for solid barriers, although it was not the expected result. This finding could be explained by the great distance between the source and the receiver. It opens the possibility of successfully using IL prediction equations for solid acoustic barriers (both thin and thick) to estimate the acoustic performance of green barriers, at least under conditions similar to those of this case study.

Keywords: noise control, green barriers, acoustic barriers, industrial noise control

1. Introduction

This case study is based on postgraduate research by Cobo Dorado [1]. It refers to a limestone calcination plant to produce powdered lime, which is located in a rural area. The industrial plant has dense forest areas in parts of its perimeter, which were planted for landscape purposes.

The main sources of noise in the plant are the crushers, the sifter, and the coal processing mill. The latter is the object of this study.

There are four receivers on the perimeter of the property where the sound pressure levels (SPL) are regularly measured; two of them, called P1 and P2, will be considered for this study.

One of the main discussions regarding barrier options to mitigate the effect of noise caused by industrial plants is whether the plant barriers such as eucalyptus plantations could effectively behave as acoustic barriers. Thus, the goal of the study was to evaluate if the presence of a eucalyptus plantation located between the coal processing mill (sound source) and P1 collaborates on reducing the SPL in P1. The main objectives were finding the best fit for the depletion law of the main noise source to better evaluate the acoustic behavior of the tree barrier comparing the accuracy of the results achieved by using four different equations to calculate the insertion loss (IL) of the green barrier under study and concluding about the possibilities of using general equations for obtaining the IL of a green barrier.

In order to carry out this research, measurements of SPL were taken, based on protocols of the National Environmental Authority. All measurements were taken with time weighting Fast at the sonometer, at a height of 1.5 m above the ground. They had a minimum duration of 15 minutes. The measurements on different days were considered.

A set of SPL measurements taken when the coal mill was the only operating source was selected. The measurements in the receivers were taken monthly. The equivalent continuous SPL (L_{eq}) was recorded each second in scale A ($L_{AF,eq}$), in scale C ($L_{CF,eq}$), and in octave bands ($L_{ZF,eq}$). The background SPL at P1 and P2 were measured during the shutdown of the plant, and their frequency spectra were also obtained.

Based on the literature review (see Section 2) and the general characteristics of the eucalyptus plantation, it appears that it could attenuate the SPL in P1.

This paper is structured in six sections. First, a theory background (Section 2) and the case study basic information (Section 3) are presented; all the relevant measured spectra are also shown in Section 3. Section 4 points out the applied methodology. Section 5 presents the calculation process; at first, the sound depletion law of the source was studied and then, the SPLs in the receiver P1 were found without studying the green barrier. For explaining the difference between the measured and calculated figures, the acoustic performance of the tree plantation was obtained four ways: two of them were for solid acoustic barriers and the other two for tree barriers. The last sections for this chapter are devoted to discussing the results and present our conclusions.

2. Background

Acoustic barriers have been widely used when noise control on the propagation path is needed, once there are no other possibilities of control on the source [2]. Up to five acoustic phenomena can occur in an acoustic barrier: sound reflection, transmission, absorption, diffraction, and scattering.

The material of the surface exposed to the source is what defines the amount of acoustic energy that will be scattered, absorbed, and reflected. When there are barriers on both sides of a source and their surface materials are not adequate for sound absorption, sound pressure levels may increase due to multiple reflections between the two sheets of the acoustic barrier [3]. The acoustic impedance of the material of the barrier determines the amount of acoustic energy that will be transmitted through it. It is generally assumed that if a material has a surface density of at least 10 kg/m^2 (kilogram per square meter), it is suitable for acting as a noise barrier [2].

Finally, diffraction should be the predominant acoustic phenomenon in a barrier. It refers to the fact that sound waves change direction, edging the obstacles they find in their path, which, in the case of a barrier, occurs at the top edge but also at the side edges [2].

The attenuation provided by an acoustic barrier is called *insertion loss* (IL), which is defined as the difference between the direct sound pressure level obtained in the absence of an acoustic barrier (L_{dir}), and the level obtained with the presence of the barrier, i.e., the diffracted sound pressure level (L_{dif}) [2] (see Eq. (1)).

$$IL = L_{dir} - L_{dif} \tag{1}$$

The proposal of Maekawa to calculate the IL value from the Fresnel number N marked a milestone in the development of noise barrier research (Eq. 2). A detailed analysis of Maekawa's work can be found in [4].

$$\mathrm{IL} = 10\ \log\ (20\ \mathrm{N}) \tag{2}$$

where $N = \frac{2\,\delta}{\lambda}$ is the Fresnel number

λ = the wavelength of sound (in meters) at the considered frequency f, and.

$\delta = a + b - d$ is the difference between diffracted sound path and direct sound path (in meters) (see **Figure 1**).

Since Maekawa's first approach had some limitations, some authors have worked on finding a better calculation method to predict the IL of acoustic barriers [2]. In next sections, some of them will be presented.

2.1 Thick barrier approach

An acoustic barrier is said to be "thick" when it has more than one point where diffraction can occur [2].

A barrier is considered thick when:

- The width of its crest exceeds 3 m. In this case, it is considered to be thick for all frequencies. Although not exactly a crest, the width of the straight line that connects the source and receiver.

- The wavelength λ to be considered is less than 1/5 of the crest width t ($\lambda < t/5$). If there were frequencies where this relation is not accomplished, the barrier would function as a thin barrier, and it should be calculated in such manner.

Eq. (3) can be used for solid barriers either thin or thick. In the case of thick barriers, the thickness t is added to the smallest of the distances a or b and with this

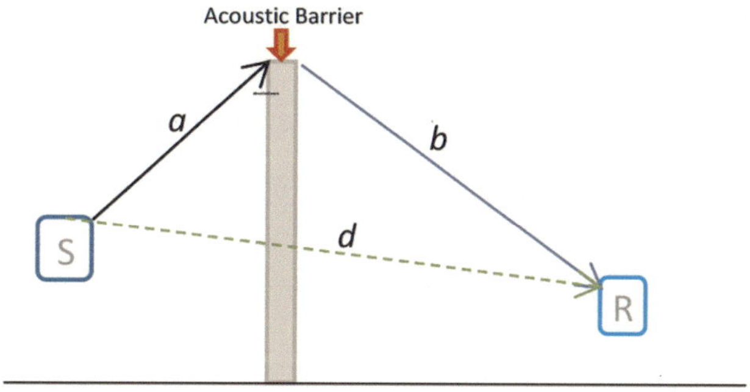

Figure 1.
Cross section of an acoustic barrier (adapted from [2]).

new value *a'* or *b'*, the Fresnel number N should be calculated. Then, the insertion loss will be obtained through Eq. (3) [2].

$$\mathrm{IL} = 10\ \log\ (3 + 10\ \mathrm{N} \cdot \mathrm{K}) - \mathrm{A_{gr}} \tag{3}$$

Where:

N: Fresnel number calculated by considering the hypothesis of thick barrier, Eq. (4) (see **Figure 2**):

$$N = \frac{2}{\lambda}(a' + b - d) \tag{4}$$

For a thick barrier, the value of *t* must be added to the minimum of *a* and *b*; the values corrected this way are noted as *a'* or *b'*.

K: meteorological correction, with.

K = 1 for distances between source and receiver either less than 100 m or greater than 300 m.

Otherwise:

$$K = e^{-0.0005\sqrt{abd/N\lambda}}$$

$\mathrm{A_{gr}}$: ground attenuation along the sound path.

The barrier attenuation should not be assumed to be greater than 20 dB.

2.2 Kurze-Anderson approach

This way of obtaining IL is a general one. The expressions to be used are those of Eq. (5) [2].

$$\text{If } \mathrm{N} > 12.5 \qquad IL = 24$$

$$\text{If } -0.2 < \mathrm{N} < 12.5\ IL = 5 + 20\ log\ \frac{\sqrt{2\pi N}}{\tanh\sqrt{2\pi N}} \tag{5}$$

Where N is the Fresnel number defined in Eq. (4)

$$\text{Remember that } tanh\,X = \frac{senh\ X}{\cosh X} = \frac{e^X - e^{-X}}{e^X + e^{-X}}$$

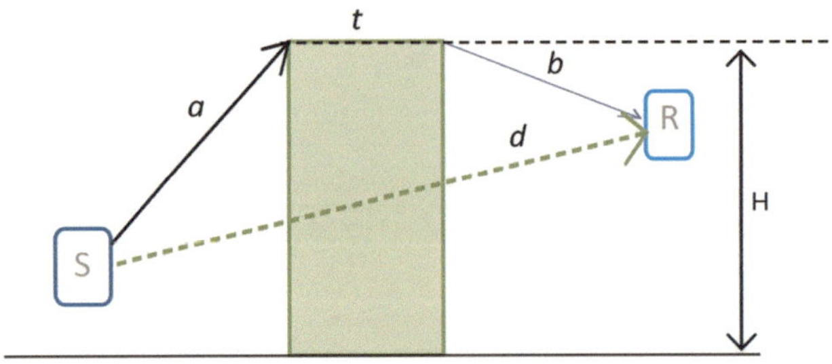

Figure 2.
Cross section of a thick barrier (adapted from [2]).

2.3 Green barriers

The case of tree or green barriers is rather different. Just as in "conventional" barriers, the hermeticity of the material is central, it is necessary to assume that the tree barriers are non-soundproof. In turn, scattering may homogenize the acoustic field into the tree plantation. On the other hand, among the characteristics of the vegetation that participate in the attenuation of the barrier, the density of the plantation, distance between trees, geometric pattern of the plantation, the features of trunks, bark, and canopy, and the dimensions of the leaves, whether the trees are deciduous or evergreen. The abovementioned points are presented in Section 2.3.1.

One simple point for expecting a good sound attenuation performance in a tree barrier is that it must block the visuals between source and receiver, as stated by ISO 9613-2 [5]. If the receiver is able to be seen from the source and through the vegetation, it is most probable that the barrier will not be dense enough for sound attenuation.

2.3.1 An overview of the research on green barriers

The acoustic of forests began to be studied in 1946 by Eyring [6], who found an attenuation of 0.05–0.13 dB/m (decibels per meter), increasing with frequency. Some years later, Embleton [7] and Aylor [8, 9] continued with Eyring's work. Embleton [7] obtained an important depletion of 7 dB/100 ft. (decibels each 100 feet) for frequencies below 2000 Hz. He tried to explain such high results thinking about branches as resonance absorbers, but his theory was not easy to be proven. Recent works, as Johansson's [10], show that the forest is a complex system that can reduce (absorb) or amplify sound pressure levels, depending on the phenomena that are activated for each case.

Aylor [8] showed that plants have a good behavior for noise control. He worked with pink noise attenuation through a dense reed marsh (*Phragmites communis*). The average height of the reeds was 2.5 m, the area of leaves per volume unit of canopy was 3.0 m^2/m^3, and the density of plants was 59 $\pm$ 10 plants/m^2. The average width of the leaves was 3.2 cm. Aylor found an increasing attenuation between 500 Hz and 10,000 Hz, with an increasing rate close to 4–4.5 dB/octave. He found an attenuation close to 18 dB at 10,000 Hz for 12.2 m broad of reeds. When comparing with corn plants (*Zea mays*) attenuation, he found that the best performance was at a frequency of 2000 Hz; the width of the corn plants leaves was 7.4 cm in average. In another study, Aylor [9] measured the sound attenuation related to trunks and stems. He used field corn (*Z. mays*), hemlock (*Tsuga canadensis*), red pine (*Pinus resinosa*), and a hardwood brush of about 6 m in height. He found the denser the plantation and the greater the leaves surface, the better the attenuation. Also, the trunks have an important effect on sound scattering, whether the wavelength was small in comparison to the radius of the trunk.

Price et al. [11] measured and studied different forests sound attenuation: Norway spruce (*Picea abies*) and oaks, with a dense undergrowth; a monoculture of Norway spruce of 11–13 m in height; a coniferous plantation including red cedar (*Thuja plicata*), Norway spruce, and Corsican pine (*Pinus nigra*). Summer and Winter measurements were performed. A general attenuation pattern was found, with a first absorption peak at about 250 Hz; a region of poor absorption performance and possible resonant amplifying, around 1000–2000 Hz; and a second attenuating region, from 2000 Hz to 10,000 Hz approximately. The authors aimed to build a predictive model, by adding the contributions to sound attenuation of ground, trunks, and foliage, calculating each one separately. On the other hand, Lee et al. [12] examined 15 sites with coniferous trees, along some roads in Virginia, USA. They concluded that only a very poor sound attenuation could be attributed to the trees. They found no differences according to the trees age, height, species, nor density at the sites.

Huddart [13] stated that noise barriers such as walls, fences, or mounds of earth are often used to reduce noise pollution from traffic, but that a tree belt would be a more environmentally friendly and esthetic option. He measured the attenuation of traffic noise through five types of vegetation up to a depth of 30 m. He verified that the foliage is important in reducing the high frequencies (above 2000 Hz), while the middle frequencies (250–500 Hz) are attenuated by the absorbent qualities of the ground. The ground absorption features can be enhanced by the roots of the plants and litter.

Huisman and Attenborough [14] showed that the acoustic response of a forest directly depends on the type of wave interference: for constructive interferences (coherent waves), sound reverberation is expected; otherwise, attenuation of the sound may occur. The authors stated that sound scattering by atmospheric turbulence is a well-known phenomenon, related with loss of coherence of waves, for wavelengths minor than the trunk diameter.

Alessandro, Barbera and Silvestrini (1987) and Stryjenski (1970), cited by Ochoa de la Torre (1999) [15], proved that the acoustic absorption of some plant species varies with the size of the leaves and the density of the foliage. Thus, noise levels decrease should only be expected for frequencies above 2000 Hz, with attenuation values of 1 dB every 10 m of depth, up to a maximum of 10 dB at 100 m or more. Furthermore, Ochoa de la Torre (1999) cites Cook and Haerbeke (1971) and Alessandro et al. (1987) that, among other conclusions, stated that: "*a screen placed close to the source is more efficient than another next to the area to be protected*"; and that "*the species to be used must be evergreen, avoiding conifers, which are the least efficient.*"

In the same direction, Tarrero (2002) [16] cites Martens and Huisman (1986), who showed that deciduous trees attenuate more than grass without trees but less than evergreen ones.

In 2002, Tunick [17] linked meteorology and sound propagation in a forest. He found a microclimate, where temperature and wind velocity are rather uniform. The main attenuation phenomena in the forest are: interfering between sound waves, both direct and reflected on the ground; scattering and absorption by ground, trunks, branches, and atmospheric turbulence. In the range of medium frequencies (250–500 Hz), ground impedance is one of the most influencing factors. For frequencies from 1000 Hz to 2000 Hz, the trunks, branches, and canopy are the main agents, acting both by sound scattering and sound absorption. For these high frequencies, these phenomena seem to have more incidence on sound attenuation than refraction effects related to the microclimate in the forest.

Martínez Sala et al. [18] carried out a study with vegetable plantations (poplars, cypresses, laurels, and orange trees), demonstrating that it is possible to improve the sound attenuation obtained from a mass of trees if their elements are ordered in a periodic way. They worked with an arrangement in regular rows, a square, rectangular, and triangular configuration of the trees. Their experimental results showed that the highest sound attenuation was obtained for a range of frequencies related to the periodicity of the array. This behavior led them to intend that these sets of trees can be seen as sonic crystals. The experimental results showed that a belt of trees organized in a periodic matrix produces attenuation peaks at low frequencies ($f < 500$ Hz), not as a consequence of the ground effect but as a result of the destructive interference of scattered waves. Therefore, these periodic arrays could be used as plant acoustic screens.

Onuu [19] found that grass can introduce an attenuation in all frequencies twice the amount of attenuation of a forest. The best performance was measured between 1000 Hz and 4000 Hz. He also stated that the best relation for representing the attenuation of grass is logarithmic, whether for trees is a power equation.

Swearingen and White [20] proposed an adjustment of the calculation method of Defrance, to include other atmospheric phenomena, especially those related to scattering. As that previous model did, they used the Green's function parabolic equation (GFPE) for modeling different phenomena that they also measured. The authors added those phenomena one by one to their simulation, and they found that the atmospheric condition had strong influence on sound propagation. Trunks and canopy scattering became more important at greater distances to the source, but they had not a significant influence on sound pressure levels, when compared to the atmospheric incidence.

In their exhaustive analysis of noise barriers, Kotzen and English [3] state that the best performance of a green barrier occurs at a frequency for which the wavelength is twice the size of the leaves of the trees or shrubs. It makes sense with Aylor's findings [8, 9], more than 30 years earlier. According to [3], green barriers are not expected to control sounds of frequencies lower than 250 Hz, and their best performance is for frequencies of 1000 Hz and higher.

Fan et al. [21] did many measurements behind six dense hedges involving six different evergreen species: arrowwood (*Vibumum odoratissimum*), oleander (*Nerium indicum*), Chinese Photinia (*Photinia serrulata*), bamboo (*Oligostachyum lubricum*), Red Robin Photinia (*Photinia fraseri*), and Deodar Cedar (*Cedrus deodara*). The authors found the best performances for the so-called "leaf shape" (the relation between leaf length to leaf width) between 2 and 3, for the greater leaf area and leaf weight: between 3 and 4 dB/m. Bamboo and oleander did not exhibit good attenuation, but deodar cedar presented very good attenuation at low frequencies (lower than 100 Hz and between 250 Hz and 800 Hz). On the other hand, both *Photinia* species and arrowwood showed their greatest attenuation at frequencies higher than 2000 Hz. Thus, the authors recommend using different kind of species in order to enhance the acoustic behavior of a hedge. They obtained Eq. (6) by regression, and they propose it for calculating the sound attenuation of hedges, in dB/m.

$$\Delta L_{Aep}\ (\mathrm{dB/m}) = 2.705 + 0.266\ W - 3.337\ T - 0.094\ S \quad (6)$$

Where:

W is the leaf weight (g)

T is the tactility; T = leaf weight/leaf area (g/cm^2)

S is leaf shape; S = leaf length/leaf width (m/m)

Horoshenkov et al. [22] demonstrated the importance of the characteristics of leaves for their acoustic performance, especially as sound absorbers. The authors worked with five kinds of plants (*Geranium zonale, Hedera helix, Pieris japonica, Summer Primula vulgaris,* and *Winter P. vulgaris*). The laboratory work was done using an impedance tube (or Kundt tube). The authors also measured the thickness, weight, and area of single leaves, the number of leaves on a plant, the volume occupied by the plant, the dominant angle of leaf orientation, the total area of leaves by plant, the surface density of a single leaf, and the total weight of leaves and stems. The *Winter Primula Vulgaris* had the best acoustic performance, with an absorption coefficient of 0.6 or greater for frequencies between 500 Hz and 1600 Hz. The lowest absorption coefficient was that of *H. helix*, with values lower than 0.2 for all frequencies lower than 1600 Hz.

According to Asdrubali et al. [23], the most important part of the attenuation in a forest is provided by the ground surface. They stated: "*the main absorber is the substrate soil (...). The presence of the plants becomes useful only when a large number of them is installed on the sample, otherwise is even pejorative within some frequency ranges.*"

On the other hand, contemporaneously, Azkorra et al. [24] obtained a weighted sound absorption coefficient of 0.40 with the best absorption behavior at frequencies of 125 and 4000 Hz, and the worst ones at 500 and 1000 Hz.

Li et al. [25] demonstrated that most of the sound absorption by trees is due to its bark properties. The rougher the surface of the bark, better sound absorption performance would be expected. When the bark had moss, the acoustic performance was significantly enhanced. In any case, the absorption coefficients for normal incidence in the range of 160–1600 Hz are actually low: the highest measured values were about 0.1, broadleaved trees having worse results than coniferous trees.

2.3.2 Hoover's expression

According to Palazzuoli and Licitra [26], the attenuation of noise traveling a distance d_f through a dense forest can be estimated using Hoover's expression (Eq. (7)).

$$A_f = \frac{d_f}{100} f^{\frac{1}{3}} \tag{7}$$

Where d_f is the distance through the forest, in meters

f is the frequency, in Hz

2.3.3 ISO 9613-2 approach

The broadly used ISO 9613-2 Standard [5] also considers the attenuation of green barriers as one of the sound attenuation terms, such as geometric divergence A_{div}, atmospheric absorption A_{atm}, ground attenuation A_{gr}, presence of obstacles A_{bar}, and miscellaneous attenuation A_{mis}. One of the miscellaneous attenuation phenomena is just the propagation through foliage A_{fol}. The general equation is Eq. (8).

$$A = A_{div} + A_{atm} + A_{gr} + A_{bar} + A_{mis} \tag{8}$$

The main terms for obtaining the sound attenuation, A_{div}, A_{atm}, and A_{gr}, are presented below. In turn, A_{fol} is also presented.

2.3.4 Acoustic divergence A_{div}

ISO 9613-2 assumes any sound source as a point one. Then, it uses a spherical or quadratic divergence, as presented in Eq. (9).

$$L_{p,r_1} = L_{p,r_0} - 10\ log\left(\frac{r_1^2}{r_0^2}\right) \tag{9}$$

Where:

r_0: distance in meters from the source where its emission sound pressure levels were measured

r_1: distance in meters from the source to the measuring point

$L_{p,r0}$: measured sound pressure level, at a distance r_0 from the source
$L_{p,r1}$: expected sound pressure level, at a distance r_1 from the source

2.3.5 Atmospheric absorption A_{atm}

The atmospheric absorption refers to the attenuation of sound due to traveling through the air along a distance d. According to [5], it should be calculated by applying Eq. (10).

$$A_{atm} = \frac{\alpha \cdot d}{1000} \tag{10}$$

Where:
d: Distance in meters from the source to the receiver
α: Atmospheric absorption coefficient, expressed in dB/km in each octave band.
α depends mostly on the frequency of the sound, the temperature, and the relative humidity of the air. Atmospheric absorption should not be greater than 15 dB at any octave band.

2.3.6 Ground attenuation A_{gr}

The attenuation A_{gr} is mostly the result of the interference between direct and reflected sound waves. This attenuation is mainly determined by the ground surface near the source and receiver. The calculation method proposed in [5] is only applicable if the terrain is flat, either horizontal or with a constant slope.

In order to calculate the attenuation due to ground absorption, three regions are defined (see **Figure 3**).

a. source region: it is the region closer to the source, in the path from the source to the receiver. It covers a length of $30.h_s$, with a maximum distance of d_p. h_s is the source height in meters, and d_p is the distance between source and receiver, in meters.

b. receiver region: it is the region closer to the receiver, in the path from the source to the receiver. It covers a length of $30.h_r$, with a maximum distance of d_p. h_r is the receiver height, in meters.

c. middle region: it is the region between the source region and the receiver region. Its length is $d_p - (30.h_s - 30.h_r)$; it is not defined when $d_p < (30.h_s + 30.h_r)$.

The total ground attenuation is obtained for each octave band, by adding the attenuations occurring in the three abovementioned zones. See Eq. (11)

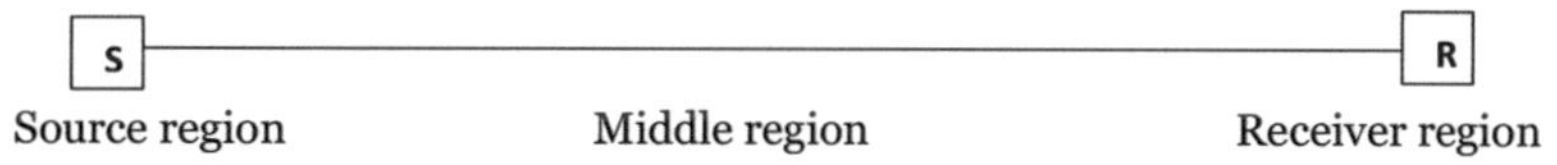

Figure 3.
Regions to calculate the attenuation due to ground absorption (based on [5]).

$$A_{gr} = A_s + A_m + A_r \tag{11}$$

Where:
A_{gr}: Total sound absorption due to ground effects (dB)
A_s: Sound absorption due to ground effects at the source region (dB)
A_m Sound absorption due to ground effects at the middle region (dB)
A_r: Sound absorption due to ground effects at the receiver region (dB)
ISO Standard 9613-2 [5] explains in detail how to calculate the values of A_s, A_m, and A_r. In the expressions for calculating the attenuation, the acoustic properties of each of these zones are taken into account through the so-called "G factor." When the sound is expected to propagate over hard ground: G = 0; for porous or soft ground: G = 1; and for mixed soil along the sound path, G should take a value between 0 and 1.

2.3.7 Foliage attenuation A_{fol}

ISO 9613-2 [5] states that the foliage of trees and shrubs provides a small amount of attenuation and only if it is sufficiently dense to completely block the view along the propagation path.

According to [5], the attenuation of sound when propagating through a green barrier or dense foliage of thickness d_f, increases with the frequency and with d_f.
The attenuation values are detailed in the Standard for d_f values less or equal to 20 m and greater than 200 m. For thicknesses from 20 m to 200 m, it may be obtained by multiplying d_f by a set of coefficients specified for each frequency band.

3. Case study basic information

As stated, the case study is based on [1]. The main characteristics of the case are presented in Section 1. In this section, it will be presented in detail.

3.1 Sound sources

There are several sound sources at the industrial plant, related to equipment, transportation, and personnel movement. **Figure 4** illustrates the location of the main sound sources at the plant.

The case study is centered only on the emissions from the main noise source in the industrial plant: the coal processing mill. All the measurements selected to work with are representative of this particular situation, since they were specially chosen, and the absence of other operating sources was checked.

The measurements for characterizing the coal processing mill were taken at 12 m distance. **Table 1** presents a representative spectrum of the source emission SPL, in octave bands. It was obtained by composing five SPL measurements, taken on different days. The obtained values were $L_{AF,eq}$ = 83.4 dB and $L_{CF,eq}$ = 84.9 dB.

3.2 Receivers

Two receivers will be considered: P1 and P2, defined as monitoring points on the property line. **Figure 5** shows their location in relation to the plant.

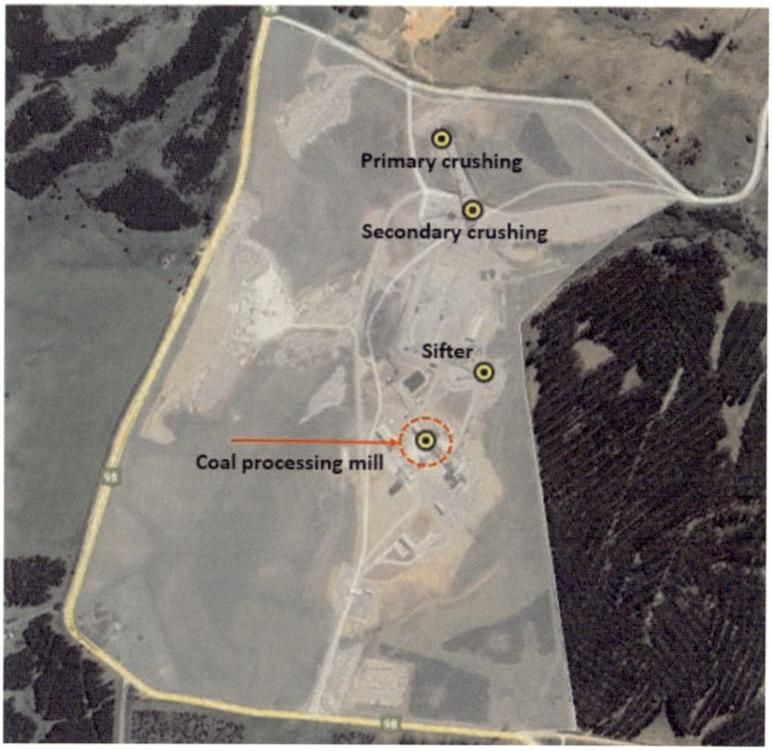

Figure 4.
Main sound sources (drawing overlapped on GoogleEarth image).

f (Hz)	31.5	63	125	250	500	1000	2000	4000	8000
$L_{ZF,12m}$ (dB)	74.5	79.6	72.6	73.1	76.3	78.0	78.3	74.9	64.8

Table 1.
Sound emission spectrum of the coal processing mill at 12 m distance.

3.2.1 Point P1

Point P1 is located facing southwest of the plant, 813 m from the coal mill. Since it is close to a neighboring house, which is temporarily occupied, it is a very important surveillance point. Our study focuses on the results at this point to assess the acoustic behavior of the green barrier.

The background noise at P1 was determined during the shutdown period of the plant. It resulted in a L_{AFeq} value of 40 dB. Its spectrum is shown in **Table 2**.

Table 3 presents the measured SPL at P1 when the coal processing mill was the only noise source operating at the plant. The measured SPL were $L_{AF,eq}$ = 47.7 dB and $L_{CF,eq}$ = 54.0 dB.

3.2.2 Point P2

Point P2 is located on the perimeter of the industrial property, with no nearby houses and close to a large external afforestation. It is located facing southeast, 239 m

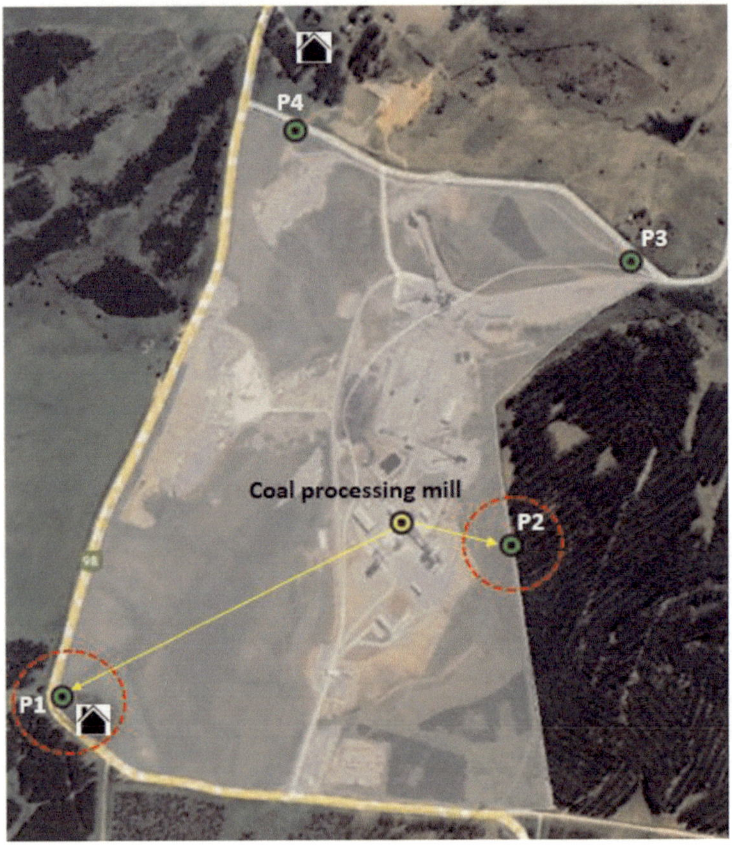

Figure 5.
Relative location of the coal processing mill, P1 and P2 (drawing overlapped on GoogleEarth image).

f (Hz)	31.5	63	125	250	500	1000	2000	4000	8000
$L_{ZF,eq}$ (dB)	50.7	49.4	41.7	35.0	35.4	34.9	30.9	32.0	24.0

Table 2.
Background noise spectrum at P1.

f (Hz)	63	125	250	500	1000	2000	4000	8000
$L_{ZF,eq}$ (dB)	52.1	43.5	45.2	45.4	40.5	40.8	38.3	30.9

Table 3.
SPL spectrum measured at P1 when the coal processing mill was the only operating noise source in the plant.

from the source. Since there are no obstacles between P2 and the coal processing mill, it is the best point to verify the sound depletion law from the source.

The background noise at P2 was also determined during the shutdown period of the plant. L_{AFeq} was 50 dB. The spectrum of background noise at P2 is shown in **Table 4**.

f (Hz)	31.5	63	125	250	500	1000	2000	4000	8000
$L_{ZF,eq}$ (dB)	56.7	59.3	49.0	44.4	45.0	44.6	43.7	39.1	28.0

Table 4.
Background noise spectrum at P2.

f (Hz)	63	125	250	500	1000	2000	4000	8000
$L_{ZF,eq}$ (dB)	67.1	58.2	59.2	60.7	57.3	55.1	49.6	37.3

Table 5.
SPL spectrum measured at P2 when the coal processing mill was the only operating noise source in the plant.

Table 5 presents the measured SPL at P2 when the coal processing mill was the only noise source operating at the plant. The measured SPL were $L_{AF,eq}$ = 62.5 dB and $L_{CF,eq}$ = 68.9 dB.

3.3 Green barrier

3.3.1 Description

The green barrier under study is *Eucalyptus dunii*, with a width of 61 m and a length of approximately 239 m, covering a total area of $1.46 \times 104\ m^2$. It is placed between the coal processing mill and Point P1, 278 m from the mill (see **Figure 6**).

E. dunii is a tree with straight trunk, with dense and drooping foliage. The projected planting technique was staggered diagonally, which is the technique commonly used from the agronomic point of view for planting such species of trees. It is based on the formation of an equilateral triangle between trees. In this case, a variant was made in such a way that the distance between trees in each row was 2.0 m and the distance between rows was 4.0 m. In any case, it can be considered that it is a regular planting pattern, as shown in **Figure 7**. Based on this configuration, there is a surface density of 0.3125 trees/m^2.

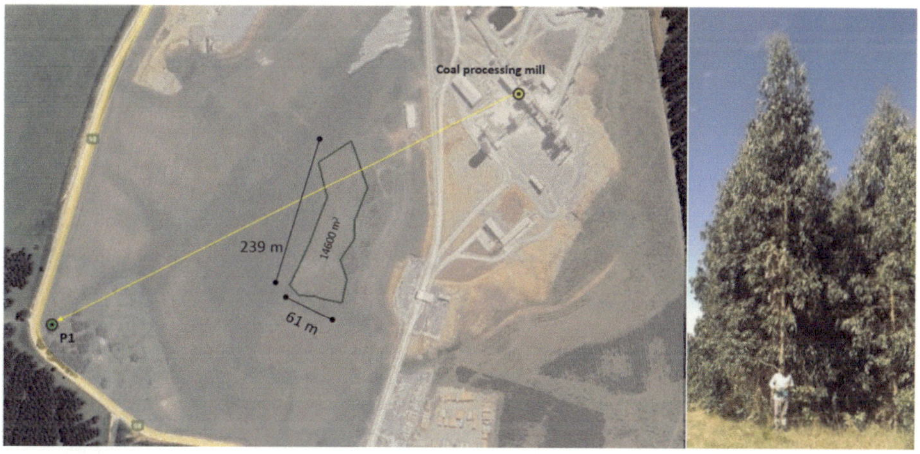

Figure 6.
Location of the tree barrier.

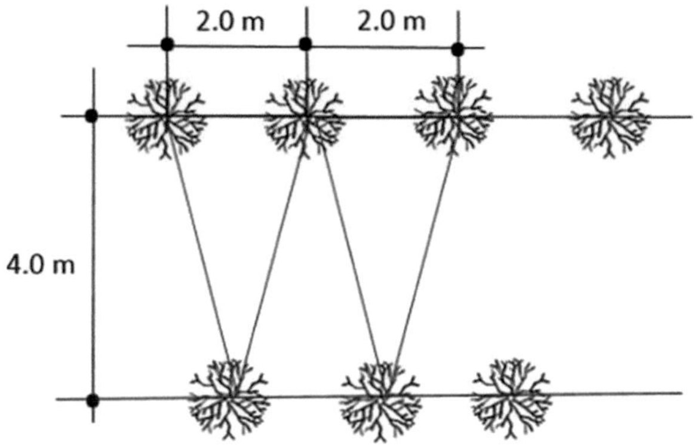

Figure 7.
Staggered planting pattern.

The height of the tree barrier was measured, and the following results were obtained:

- Shortest trees: 8 m
- Tallest trees: 11.3 m
- Average height of the barrier: 10.4 m
- Perimeter of the trunk at shoulder height: 61 cm
- Average length of the leaves: 22 cm

Since the plantation was not created for commercial use, no pruning of root sprouts was done. This practice increases the density of the barrier.

General recommendations	Tree barrier situation
Green barriers are not suitable for controlling low frequency noise [13]	The measured sound pressure levels show that C-A is approximately 2 dB. Thus, the mill sound emissions have no significant low frequency components.
The evergreen trees/shrubs have better acoustic performance than the deciduous ones [16]	It is evergreen eucalyptus. Since they are not for forest use, they are not pruned.
Every 10 m in depth (dense vegetation), approximately 1 dB reduction is achieved [15]	The tree barrier is 61 m thick in the line that joins the source and the receiver. In-depth attenuation of about 6 dB is possible.
The barrier will be more effective when closer to the source [15]	The tree barrier is 278 m from the source and 474 m from the point of measurement.

Low-frequency noise has significant acoustic energy in the frequency range from 20 Hz to 100 Hz. The comparison of the sound level on scale C with the sound level A, allows determining whether or not there are significant low-frequency components. Indeed, since curve A attenuates low frequencies and curve C does not, if the difference between dBA and dBC values is not huge (C-A $\leq$ 10), it will be considered that the low frequency components are not important [27].

Table 6.
Tree barrier initial check.

3.3.2 Initial check

At first, some general conditions were checked to know if the tree plantation would have a noticeable acoustic performance (**Table 6**).

The tree barrier features make possible to expect some sound pressure levels reduction at P1.

4. Methodology

This study is based on postgraduate research by Cobo Dorado [1]. A large set of SPL measurements were performed, using a CESVA SC-30 Sound Level Meter with Octave Band Analyzer, Type 1 according to IEC 60651:01 and 60,804:00. Calibration was checked before and after each measurement with a portable calibrator CESVA CB006, Type 1. A windscreen was used in all cases. The instrument was placed on a tripod at a height of 1.50 m.

The meteorological variables were measured at each measurement point using a portable climatic station with a digital anemometer Speedtech WindMate 300. A fixed meteorological station located at the perimeter of the industrial plant was used as a local reference element. This station records every 10 minutes, date, time, wind speed (m/s), wind direction (degrees), temperature (°C), and relative humidity (%) and air quality parameters.

A Garmin GPS was used for georeferencing of each measurement point.

To carry out the assessment of the acoustic performance of the tree barrier under study, the following approach has been used:

1. Based on the SPL measured values in the source and the receiver P2, the behavior of the source was analyzed to find the depletion law of sound pressure levels. The objective was determining if the geometric divergence law of the coal processing mill fitted a spherical or a cylindric depletion law.

2. The expected sound pressure levels caused by the coal processing mill at P1 were calculated, only considering the geometric divergence law obtained in the previous step, the atmospheric absorption and the ground attenuation, i.e., the direct propagation without considering the presence of the vegetation barrier.

3. Results were compared with those measured in P1, focusing on the possibility of an extra attenuation due to the presence of the green barrier.

4. The expected SPL from the coal mill at P1 were calculated again, adding the attenuation of the green barrier to the sound pressure levels calculated above. Four different equations were used to estimate the barrier attenuation. When IL equations for solid barriers were used, the diffractions above and along the sides of the barrier were also considered.

5. Results were compared again with those measured in P1.

6. Conclusions about the acoustic performance of the tree barrier and which are the better equations to predict it were found.

All the data processing was performed using electronic spreadsheets.

5. Acoustic performance of the green barrier

5.1 Finding the depletion law of the noise source

The source's emission spectrum measured at 12 m was propagated to obtain the expected spectrum at P2, considering A_{div} and A_{atm}. For calculating A_{div}, both spherical and cylindrical propagations were tested, i.e., applying Eqs. (9) and (12), respectively [1].

$$L_{p,r_1} = L_{p,r_0} - 10\ log\left(\frac{r_1}{r_0}\right) \tag{12}$$

Where

r_0: distance from the source where emission sound pressure levels were measured (12 m)

r_1: distance from the source to P2 (239 m)

$L_{p,r0}$: measured sound pressure level, at a distance r_0 from the source (see **Table 1**)

$L_{p,r1}$: expected sound pressure level, at a distance r_1 (in this case, at P2)

Since the distance from the source to P2 is further than 100 m, the atmospheric absorption will also be considered for this comparison. The atmospheric absorption refers to the attenuation of sound due to traveling along a distance *d*. According to [5], it should be calculated by applying Eq. (10).

In this case, the average temperature and humidity conditions at P2 during the sound pressure level measurements (**Table 5**) were T = 25°C and RH = 50%. The values of α in **Table 7** were taken from Miyara [28].

The results are shown in **Table 8** and **Figure 8**. As it can be seen in **Figure 8**, the linear approach fits better the measured values up to 500 Hz. Thus, the calculation method was not that of ISO 9613-2 [5] because the divergence calculation law was assumed to be not quadratic but linear. All the other attenuation terms were calculated according to ISO 9613-2.

5.2 Sound pressure levels at P1, excluding the tree barrier acoustic performance

SPL at P1 were calculated by using Eq. (12). In this case, r_0 was 12 m and r_1 was 813 m.

f (Hz)	63	125	250	500	1000	2000	4000	8000
α (dB/100 m)	—	0.04	0.13	0.35	0.70	1.18	2.52	7.59

Table 7.
Atmospheric absorption coefficients by octave bands for T = 25°C and HR = 50% (from [28]).

f (Hz)	63	125	250	500	1000	2000	4000	8000	$L_{AF,eq}$
L_{ZF} measured	67.1	58.2	59.2	60.7	57.3	55.1	49.6	37.3	62.5
L_{ZF} calculated with Eq. (9) (quadratic divergence)	53.6	46.5	46.8	49.5	50.4	49.5	42.9	23.8	54.8
L_{ZF} calculated with Eq. (12) (linear divergence)	66.6	59.5	59.8	62.5	63.3	62.5	55.9	36.8	67.8

Table 8.
Comparison of sound pressure levels at P2 using two different depletion laws (all values are in dB).

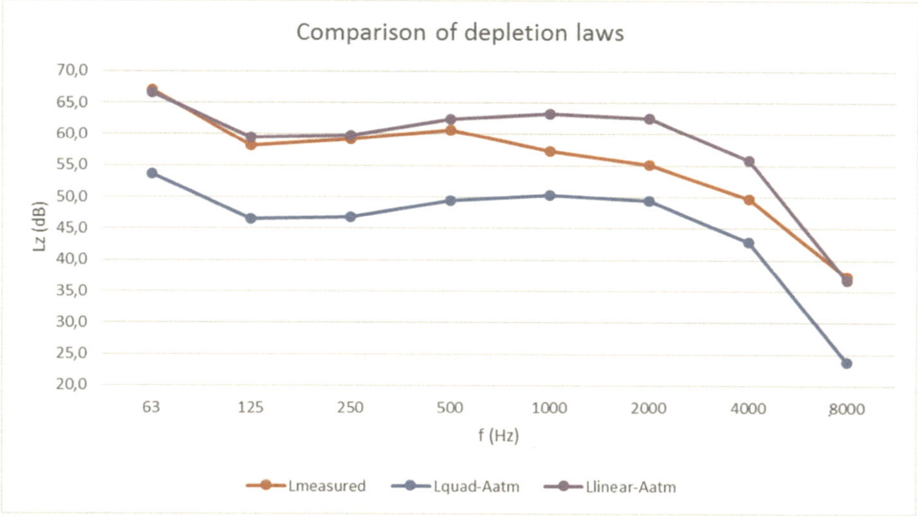

Figure 8.
Comparison of spectra obtained with linear and quadratic depletion laws and the measured spectrum.

Atmospheric and ground absorptions were also considered. Since the ground slope between the coal processing mill and Point P1 is rather uniform (**Figure 9**), the calculation approach proposed by ISO 9613-2 [5] for A_{gr} can be used.

The SPL at P1 were measured when the coal mill was the only noise source operating in the plant. Meteorological conditions during the measurements were considered in selecting atmospheric absorption coefficients (T = 20°C and RH = 80%) (see **Table 9**).

The sound path between the mill and P1 consists of various types of soil, as sketched in **Figure 10**. The length and type of surface of each one are summarized in **Table 10**. The detailed method for calculating G can be found at [5].

Taking into account the height of the source h_s = 3.6 m and the height of the receiver in P1 h_r = 1.5 m, the values of G for each region in **Figure 3** are presented in **Table 11**.

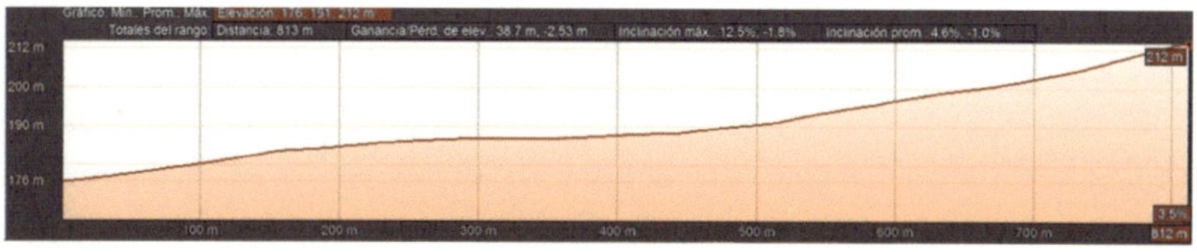

Figure 9.
Diagram of the terrain profile from the coal processing mill to P1 (obtained from Google Earth).

	63	125	250	500	1000	2000	4000	8000
α (dB/100 m)	—	0.03	0.10	0.28	0.52	0.90	2.13	6.86

Table 9.
Atmospheric absorption coefficients by octave bands for T = 20°C and HR = 80% (from [28]).

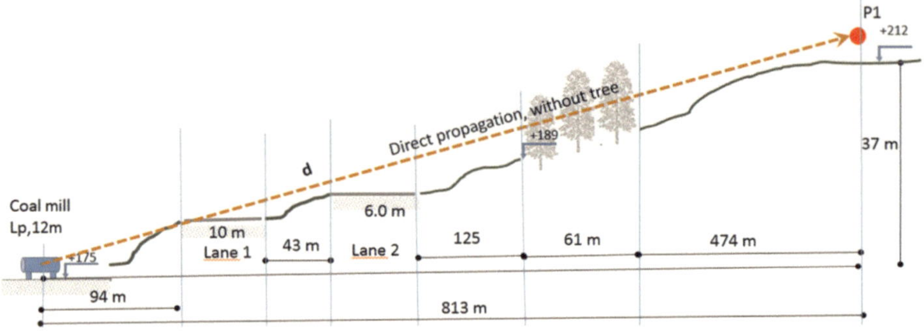

Figure 10.
Diagram of the case study (not in scale).

From	to	Distance (m)	Soil type
Coal mill	End of the mill base	5	Concrete
End of the mill base	Road lane 1	89	Compacted soil
Lower edge of road lane 1	Upper edge of road lane 1	10	Hard pavement
Upper edge of road lane 1	Lower edge of road lane 2	43	Soil
Lower edge of road lane 2	Upper edge of road lane 2	6	Hard pavement
Upper edge of road lane 2	Lower edge of green barrier	125	Soil/Grass
Lower edge of green barrier	Upper edge of green barrier	61	*Eucalyptus dunii*
Upper edge of green barrier	Point P1	474	Soil/Grass

Table 10.
Ground absorption characteristics.

	Source region	Intermediate region	Receiver region
Length	108 m	660 m	45 m
Ground coverage	Concrete/compacted soil/hard pavement/soil	Soil/grass/trees/hard pavement	Soil/grass
G factor	$G_s = 0.2$	$G_m = 0.9$	$G_r = 1$

Table 11.
G values and ground coverage, by region.

The propagation from the sound source, only considering attenuation by distance, by atmospheric absorption, and by absorption from the ground, leads to the results in **Table 12** (calculations were done according to [5]). The 31.5 Hz band was not used, because the atmospheric absorption coefficient does not calculate the same way as it does in higher frequencies.

When comparing the results in the first and the last row in **Table 12**, it appears that the measured values are lower than the ones previously calculated, when expressed in A-weighting scale. The difference is greater at the lowest frequency band and at 1000 Hz and 2000 Hz bands.

According to the background discussed in Section 2, these differences reinforce the hypothesis of an extra sound attenuation, possibly provided by the tree barrier.

f (Hz)	63	125	250	500	1000	2000	4000	8000	$L_{A,eq}$
L_z measured	52.1	43.5	45.2	45.4	40.5	40.8	38.3	30.9	47.7
$L_{Z,813m}$ (Eq. 12)	61.3	54.3	54.8	58.0	59.7	60.0	56.6	46.5	
A_{atm}	0.0	0.2	0.9	2.3	4.2	7.3	15.0	15.0	
A_{gr}	−5.4	3.9	6.1	3.5	−0.8	−1.4	−1.4	−1.4	
$L_{Z,direct}$ at P1	66.7	50.1	47.8	52.2	56.3	54.1	43.0	32.9	59.5

Table 12.
Expected results for direct propagation, without considering the tree barrier (all values are in dB).

Tunick [17] states that the trunks, branches, and crowns are the main agents that attenuate sounds from 1000 Hz to 2000 Hz. On the other hand, according to Martínez Sala [18], the attenuation in frequencies lower than 500 Hz is due to the destructive interference of the sound waves when scattered in a belt of trees planted following a periodic pattern.

5.3 SPL at P1, considering the tree barrier insertion loss (IL)

The IL can be calculated using different formulae. Once the direct sound pressure levels L_{dir} have been calculated considering A_{div}, A_{atm} and A_{gr} (see **Table 12**), the sound pressure levels L_{dif} can be also obtained by difference, using Eq. (1). It is assumed that the SPL at the receiver were only caused by the sound wave diffracted by the barrier, i.e., no direct sound from the source was expected to arrive to P1. It also must be taken into account that there is diffraction by the lateral edges, which must be calculated and added to the previously calculated SPL at the receiver.

In this case study, IL will be calculated according to different methods, to compare their results. The approaches to be considered are: Kurze-Anderson and thick barrier approach, which are general expressions for solid barriers; and A_{fol} from ISO 9613-2 and Hoover's expression, which are specific approaches for green barriers [1, 2, 5, 26].

5.3.1 Kurze-Anderson approach

This way of obtaining IL is a general one; it has not been developed for green barriers. It is expected to overestimate the value of IL.

For a thick barrier, the value of t (**Figure 2**) must be added to the minimum of a and b. In this case, since $b > a$, then $a' = a + t$.

Thus, a = 278.08 m; b = 474.08 m; t = 61 m; a' = 339.08 m; d = 813.00 m.

The IL calculated using Eq. (5) and the SPL expected at the receiver are presented in **Table 13**. Note that the IL for the band of 63 Hz has been considered because the wavelength at this frequency is significantly shorter than the barrier width t.

The calculated SPL at 4000 Hz and 8000 Hz were lower than the background noise at P1; thus, they have been replaced by the background values in **Table 2** (figures in green in **Table 13**).

5.3.2 Thick barrier approach

The verification for this case study (t = 61 m) is presented in **Table 14**.

f (Hz)	63	125	250	500	1000	2000	4000	8000	$L_{A,eq}$
L measured	52.1	43.5	45.2	45.4	40.5	40.8	38.3	30.9	47.7
L_{dir} without tree barrier	66.7	50.1	47.8	52.2	56.3	54.1	43.0	32.9	59.5
IL K-A (Eq. (5))	6.0	6.9	8.3	10.3	12.9	15.8	18.8	21.8	
Expected SPL at P1 (L_{dif})	60.7	43.2	39.6	41.9	43.4	38.3	32	24	46.5

Table 13.
IL according to Kurze-Anderson and expected sound pressure levels at P1 (all values are in dB).

f (Hz)	31.5	63	125	250	500	1000	2000	4000	8000
λ (m)	10.92	5.46	2.75	1.38	0.69	0.34	0.17	0.09	0.04
1/5 t (m)	12.2	12.2	12.2	12.2	12.2	12.2	12.2	12.2	12.2

Table 14.
Verification of thick/thin barrier criteria (all values are in m).

Based on these results, the green barrier could be considered a thick barrier for all the frequencies. Eq. (3) can be used for solid barriers either thin or thick. Since it is not developed for tree or green barriers, overestimation of IL is expected.

The IL calculated using Eq. (3) without subtracting A_{gr} and the SPL expected at the receiver are presented in **Table 15**. Since the distance between the source and P1 is greater than 300 m, K = 1. A_{gr} has just been considered by the general attenuation terms; it must not be subtracted twice.

Note that the IL for the band of 63 Hz has been considered because the wavelength at this frequency is significantly shorter than the barrier width t.

The calculated SPL at 4000 Hz and 8000 Hz were lower than the background noise at P1; thus, they have been replaced by the background values in **Table 2** (figures in green in **Table 15**).

5.3.3 Hoover's expression

Hoover's expression has been presented in section 2, Eq. (7). It depends on the thickness of the green barrier and the frequency of sound.

The IL calculated using Eq. (7) and the SPL expected at the receiver are presented in **Table 16**. In this case, as the main considered phenomenon is the attenuation by the leaves and canopy, attenuation at 63 Hz would not be considered [3].

f (Hz)	63	125	250	500	1000	2000	4000	8000	$L_{A,eq}$
L measured	52.1	43.5	45.2	45.4	40.5	40.8	38.3	30.9	47.7
L_{direct} without tree barrier	66.7	50.1	47.8	52.2	56.3	54.1	43.0	32.9	59.5
IL TB (Eq. (3)) without A_{gr} term	5.6	6.2	7.3	8.9	11.0	13.4	16.1	19.0	
Expected SPL at P1 (L_{dif})	61.1	43.9	40.5	43.3	45.3	40.7	32	24	48.3

Table 15.
IL according to thick barrier approach and expected sound pressure levels at P1 (all values are in dB).

f (Hz)	63	125	250	500	1000	2000	4000	8000	$L_{A,eq}$
L measured	52.1	43.5	45.2	45.4	40.5	40.8	38.3	30.9	47.7
L_{direct} without tree barrier	66.7	50.1	47.8	52.2	56.3	54.1	43.0	32.9	59.5
IL H (Eq. (7))	0.0	3.1	3.8	4.8	6.1	7.7	9.7	12.2	
Expected SPL at P1	66.7	47.0	44.0	47.4	50.2	46.4	33.3	24	53.2

Table 16.
IL according to Hoover's expression and expected sound pressure levels at P1 (all values are in dB).

The only correction needed was that of background noise at 8000 Hz, in order to avoid a calculated value lower than the measured one.

5.3.4 A_{fol} *from ISO 9613-2 approach*

For this case study, d_f = 61 m. Thus, according to ISO 9613-2, the attenuation values to be considered are presented in **Table 17**.

The IL due to the propagation through a green barrier or dense foliage according to ISO 9613-2 and the SPL expected at the receiver are presented in **Table 18**. No values were needed to be replaced.

5.4 Edge diffraction

When an acoustic barrier is calculated as a "conventional" one (e.g., by using Kurze-Anderson's approach or thick barrier approach), not only the top edge diffraction is to be considered, but also the diffraction at its sides. The SPL due to the diffraction at its sides should be added to the expected sound pressure levels at the receiver.

Figure 11 shows a diagram of the sides that should be considered to obtain the diffracted SPL at the receiver. Since lateral paths are not symmetric, each of them will be calculated separately.

f (Hz)	63	125	250	500	1000	2000	4000	8000
A_f for 20 m $\leq d_f \leq$ 200 m	0.02	0.03	0.04	0.05	0.06	0.08	0.09	0.12

Table 17.
Green barrier attenuation, in dB/m (from [5]).

f (Hz)	63	125	250	500	1000	2000	4000	8000	$L_{A,eq}$
L measured	52.1	43.5	45.2	45.4	40.5	40.8	38.3	30.9	47.7
L_{dir} without tree barrier	66.7	50.1	47.8	52.2	56.3	54.1	43.0	32.9	59.5
IL ISO	1.2	1.8	2.4	3.1	3.7	4.9	5.5	7.3	
Expected SPL at P1 (L_{dif})	65.5	48.3	45.4	49.1	52.6	49.2	37.5	25.6	55.5

Table 18.
IL according to ISO 9613-2 and expected sound pressure levels at P1 (all values are in dB).

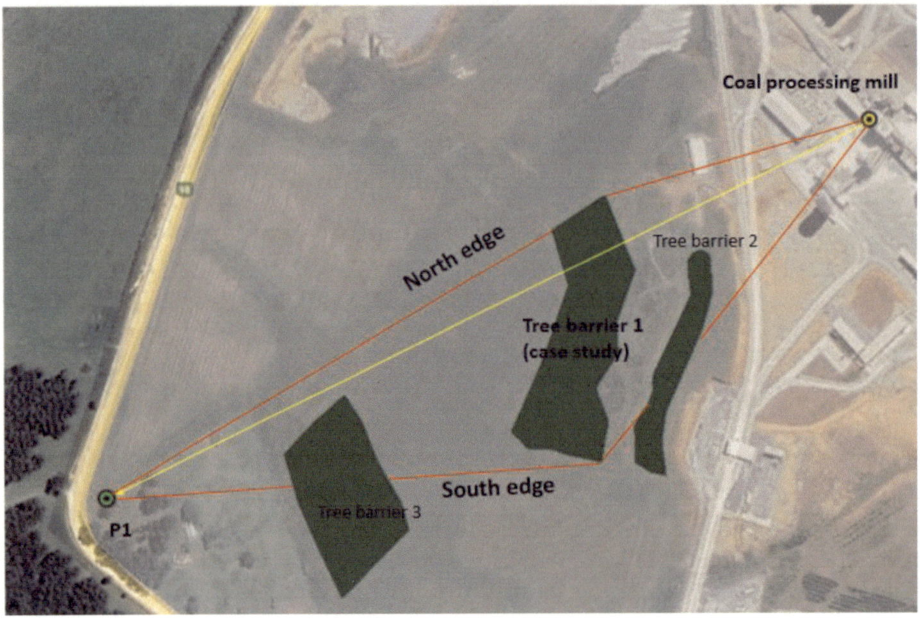

Figure 11.
Diffraction lateral paths to be considered.

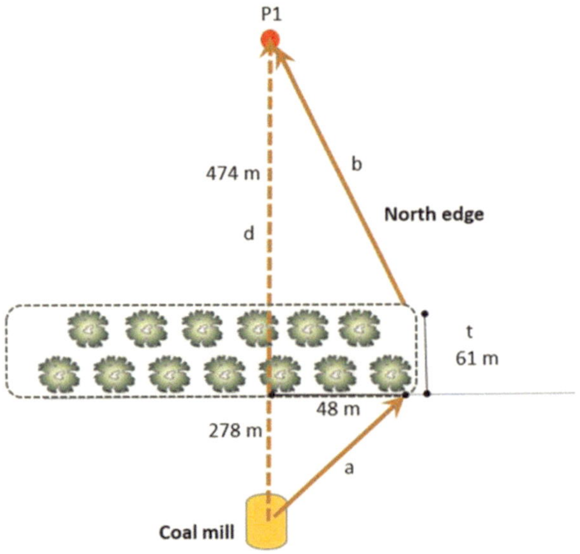

Figure 12.
Diagram of dimensions for calculation of North side diffraction.

5.4.1 North side diffraction

Figure 12 presents the diagram to be used for calculating the North side diffraction. When Kurze-Anderson or the thick barrier approach is used, the calculated sound pressure levels at the receiver were lower than the measured background noise. Thus, they are negligible.

5.4.2 South side diffraction

Figure 11 shows that before the sound reaches the South side of the barrier, it must pass through another plantation of trees along 110 m. After reaching the barrier under study, it has to pass through another forestation, in order to reach the receiver. This complex path may impose greater attenuation than a single tree barrier. Even if there are no simplified methods for calculating the SPL at the receiver in this case, the low SPL obtained at the North side allows to expect negligible results at the receiver.

5.5 Comparison of the results achieved by different calculation methods

Figure 13 shows the SPL in octave bands obtained for different situations and calculation methods.

The blue lines represent the SPL measured at P1: the bottom dotted line corresponds to the background SPL, and the solid one represents the SPL measured when the coal processing mill was the only operating source at the industrial plant.

The upper dotted line represents the calculated SPL without considering the tree barrier (L_{dir}).

All the other lines correspond to different calculation approaches. The green ones were obtained by calculation with methods that consider green barriers (Hoover's and A_{fol} from ISO 9613-2); the red and orange ones correspond to the calculation methods for solid barriers.

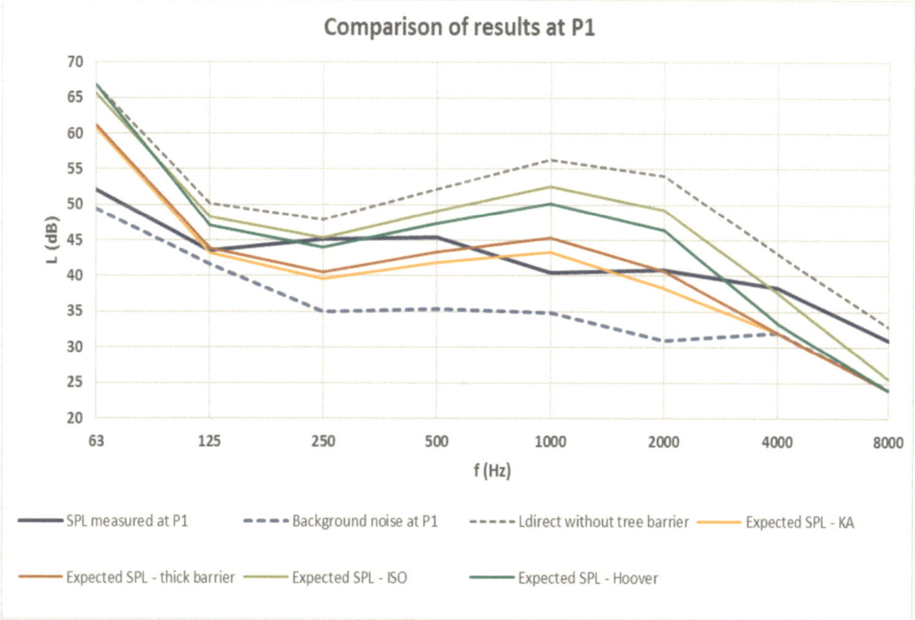

Figure 13.
Comparison of results at P1.

6. Results and discussion

The values of SPL at P2 obtained by measurement and by propagation from the source with two different depletion laws showed that the best fitting was obtained when considering a linear depletion. We focused our comparison on frequencies up to 500 Hz, because at upper frequencies the differences could be related to other phenomena.

When using an approach similar to ISO 9613-2 to propagate the SPL from the source to P1 without considering the eucalyptus plantation, a 12 dB difference was found. This difference could be due to the presence of the trees, which could behave as an acoustic barrier. The greatest extra attenuation was obtained at the frequencies of 1000 Hz and 2000 Hz; according to Tunick [17], this is the frequency range where the trunks, branches, and crowns have their best acoustic performance. Some extra attenuation was also found below 500 Hz; according to Martínez-Sala [18], the differences in this range would be attributed to the destructive interference of the sound waves when scattered in a belt of trees planted following a periodic pattern.

For answering the question about which are the best equations to predict the behavior of the vegetal barrier, the following results are discussed.

The best performance was expected for those formulae developed for green barriers, as Hoover's or the ISO 9613-2 correction term for green barriers. But when calculations were done, they achieved the worst results, being ISO 9613-2 worse than Hoover's (see **Figure 13**).

Just the opposite, the best result for the green barrier IL was achieved in the thick barrier approach, and the second in accuracy was the Kurze-Anderson approach.

In both cases, when adding the edge diffraction, the results did not exhibit any changes, i.e., the edge diffraction was significantly lower than the upper one.

It is noticeable that the results at the frequencies where green barriers are expected to have better performance (1000 Hz and 2000 Hz, according to [3] and [17]) are particularly accurate in both cases. Since these methods have not been developed for green barriers, we did not expect these results.

In order to explain these results, we think that it is possible that the sound waves could behave as if the barrier was a solid obstacle, regarding the long distance between the source and the receiver. Since our atmospheric measurement conditions according to Pasquill-Gifford (see, e.g., [29]) were unstable or neutral atmosphere (wind velocities lower than 5 m/s, variable insolation conditions), this interpretation could oppose [20], by assigning no importance to atmospheric stability conditions.

7. Final remarks

A case study about the acoustic performance of a tree barrier of *Eucaliptus dunii* has been presented. Some different approaches were used in order to calculate its insertion loss *IL*.

The noise source shows a linear SPL depletion law.

Also, the green barrier is performing as an acoustic one.

When comparing the data measured in P1 and the values calculated with direct propagation but without considering the presence of trees, an extra attenuation of approximately 12 dB with A-weighting appeared. The differences in frequencies of 1000 Hz and 2000 Hz wer even greater: 15 dB and 13 dB, respectively. For frequencies of 500 Hz and lower, there are also differences but not as huge as the abovementioned ones. Since this sound attenuation could be intended as due to the presence of the tree barrier, it confirms the hypothesis that green barriers can behave as acoustic ones.

Four options were tried for calculating the sound attenuation provided by the green barrier. Kurze-Anderson and the thick barrier approach gave a good prediction of SPL at the receiver, both in octave bands and for A-weighted values, especially for frequencies between 1000 Hz and 4000 Hz, where the results without considering the tree barrier attenuation were the least accurate. Adding the lateral diffraction did not improve the calculated results in this case.

The best approach for calculating the green barrier *IL* was the thick barrier approach and the second in accuracy was Kurze-Anderson.

The approach of ISO 9613:2 was the least accurate, worse than Hoover's approach.

It is possible that the long distance between the source and the receiver—and also the long distance between the source and the green barrier, makes the barrier behave as if it was a solid obstacle, while the leaves and canopy effects become negligible.

This finding opens the possibility of successfully using the IL prediction equations for solid acoustic barriers (both thin and thick) to estimate the acoustic performance of green barriers, at least under conditions similar to those of this case study. Further research should be needed to recommend wider use, e.g., for distances further than a given one.

Author details

Martha Cobo Dorado, Gissell Rodríguez Milan and Alice Elizabeth González*
University of the Republic, UdelaR (Universidad de la República), Montevideo, Uruguay

*Address all correspondence to: elizabet@fing.edu.uy

References

[1] Dorado C, María M. Influencia de la vegetación en la mitigación del ruido. Estudio de una forestación. Universidad de la República, Uruguay: Magister Thesis on Environmental Engineering; 2022

[2] González AE. Overview of noise control techniques and methods. In: Caniato M, editor. Noise Control. London, UK, UK: IntechOpen; 2022

[3] Kotzen B, English C. Environmental Noise Barriers: A Guide to Their Acoustic and Visual Design. 2nd ed. Canadá, USA: Simoultaneouly Published; 2009

[4] Menounou P. A correction to Maekawa's curve for the insertion loss behind barriers. The Journal of the Acoustical Society of America. 2001, 2001;**110**:1828-1838

[5] International Standard Organization. ISO 9613- Acoustics — Attenuation of sound during propagation outdoors — Part 2: General method of calculation. 1996

[6] Eyring CF. Jungle acoustics. The Journal of the Acoustical Society of America. 1946;**18**(2):257-270

[7] Embleton T. Sound propagation in homogeneous deciduous and Evergreen woods. The Journal of the Acoustical Society of America;**35**:1119-1125

[8] Aylor D. Sound transmission through vegetation in relation to leaf area density, leaf width, and Breadth of Canopy. The Journal of the Acoustical Society of America. 1972;**51**:411-414

[9] Aylor D. Noise reduction by vegetation and ground. The Journal of the Acoustical Society of America. 1972, 1972;**51**:197-205

[10] Johansson E. The Sound Amplifyng Forest, with Emphasis on Sounds from Wind Turbines. Göteborg, Sweden: Chalmers University of Technology; 2010

[11] Price Margaret A, Attenborough K, Heap Nicholas W. Sound attenuation through trees: Measurements and models. The Journal of the Acoustical Society of America. 1988;**84**:1836-1844

[12] Lee SE, Velasquez S, Flintsch G, Peterson J. Road Noise Attenuation Study: Traffic Noise, Trees, and Quiet Pavement: A Report in Response to item 442 c. of Chapter 847 of the Acts of Assembly of 2007. USA: The Virginia Transportation Research Council; 2008

[13] Huddart L. The Use of Vegetation for Traffic Noise Screening. United Kingdom; 1990

[14] Huisman WHT, Attenborough K. Reverberation and attenuation in a pine forest. The Journal of the Acoustical Society of America. 1991;**90**:2664

[15] Ochoa de la Torre. La vegetación como instrumento para el control microclimático en línea. Tesis Doctoral. Universidad Politécnica de Cataluña consultado en agosto de 2022

[16] Tarrero AI.. Propagación del sonido en bosques. Análisis comparativo de las medidas in situ, en laboratorio y de los valores predichos por un modelo en línea. Tesis Doctoral. Universidad de Valladolid consultado en agosto de 2022. 2002. https://www.cervantesvirtual.com/obra/propagacion-del-sonido-en-bosques-analisis-comparativo-de-las-medidas-in-situ-en-laboratorio-y-de-los-valores-predichos-por-un-modelo–0/

[17] Tunick A. Coupling Meteorology to Acoustics in Forests. Adelphi, MD: U.S. Army Research Laboratory; 2002

[18] Martínez-Sala R, Rubio C, Sánchez Pérez EA, Linares J. Control of noise by trees arranged like sonic crystals. Journal of Sound and Vibration. 2006;**291**:100

[19] Onuu MU. Modelling of excess noise attenuation by grass and forest. Nigerian Journal of Physics. 2006;**18**:197-202

[20] Swearingen ME, White MJ. Influence of scattering, atmospheric refraction, and ground effect on sound propagation through a pine forest. The Journal of the Acoustical Society of America. 2007;**122**:113

[21] Yang F, Bao Z, Zhu Z, Jiani L. The investigation of noise attenuation by plants and the corresponding noise-reducing spectrum. Journal of Environmental Health. 2010;**72**:8

[22] Horoshenkov Kirill V, Khan A, Benkreira H. Acoustic properties of low growing plants. The Journal of the Acoustical Society of America. 2013;**133**: 2554

[23] Asdrubali F, D'Alessandro F, Mencarelli N, Horoshenkov KV. Sound absorption properties of tropical plants for indoor applications. In: The 21st International Congress on Sound and Vibration. Beijing/China; 2014

[24] Azkorra Z, Pérez G, Coma J, Cabeza LF, Bures S, Álvaro JE, et al. Evaluation of green walls as a passive acoustic insulation system for buildings. Applied Acoustics. 2015;**89**:46-56

[25] Li M, Renterghem V, Timothy, Kang J, Verheyen K, Botteldooren D. Sound absorption by tree bark. Applied Acoustics. 2020;**165**:9

[26] Palazzuoli D, Licitra G. Chapter 1 fundamentals. In: Licitra G, editor. Noise Mapping in the EU, Models and Procedures. Boca Raton, USA: CRC Press; 2013. p. 412

[27] Boletín Oficial del Estado Español. Real Decreto 1367/2007, de 19 de octubre, por el que se desarrolla la Ley 37/2003, de 17 de noviembre, del Ruido, en lo referente a zonificación acústica, objetivos de calidad y emisiones acústicas. BOE núm. 254, de 3/10/2007. 2007

[28] Miyara F. Control de ruido. Rosario, Argentina: Own Electronic Edition; 1999

[29] Wark K, Warner CF. Contaminación del aire: origen y control. México: LIMUSA; 1990

Chapter 5

Infrasound Exposure: High-Resolution Measurements Near Wind Power Plants

Huub Bakker, Mariana Alves-Pereira, Richard Mann, Rachel Summers and Philip Dickinson

Abstract

This chapter focuses on infrasonic (≤20 Hz) noise exposure as captured in and around homes located in the vicinity of wind power plants. Despite persistent noise complaints by local residents, no satisfactory acoustical event has yet been identified to justify this troublesome (worldwide) situation. Continuous (days), high-resolution recordings—spectral segmentation of 1/36 of an octave and 1-second temporal increments—have been acquired in many homes across the world revealing the presence of wind turbine acoustic signatures. These consist of trains of airborne pressure pulses, identified in the frequency domain as harmonic series with the fundamental frequency equal to that of the blade-pass frequency of the wind turbine. This report documents three such cases (Portugal and Scotland). The highest peaks of the wind turbine acoustic signature (up to 25 dB over background noise) occurred within the 0.5–5 Hz window which is classically defined as below the human hearing threshold; and yet these 'inaudible' phenomena appear to trigger severe biological reactions. Based on the prominence of the peaks in the harmonic series, a new measure is proposed for use in determining dose–response relationships for infrasonic exposures. This new methodology may be applicable to infrasonic exposures in both environmental and occupational settings.

Keywords: harmonic series, harmonic prominence, wind rose, 1/36-octave bandwidth, time profile, low frequency noise, environmental noise, wind turbines

1. Introduction

Hearing loss, speech intelligibility and noise annoyance are some of the most studied impacts of noise exposures on human health and well-being. A common denominator of these three outcomes is the audibility of the sound. Exposure to loud noise over extended periods of time can cause hearing impairment; noisy environments can interfere with the correct understanding of speech; and certain types of continuous or intermittent sounds can cause people to feel annoyed by noise, which can, in turn, exacerbate underlying disorders or diseases.

IntechOpen

There are, however, additional features of sonic environments that are unrelated to the human audibility of sound, but that can also deleteriously affect human health and well-being, specifically, infrasound (≤20 Hz).

1.1 Infrasound and human health: brief overview

With the growing industrialization and mechanization that occurred worldwide in the 1960s, infrasound in the environment began to take its toll on workers and urban citizens. Thus, in 1973, the National Research Council of France organized an International Colloquium entirely dedicated to infrasound [1]. One of the outcomes was the establishment of permissible levels for infrasound exposures in the Russian Federation [2]. **Figure 1** shows the legislated values for the year 2000.

With the introduction of industrial wind turbines (IWT) in mostly rural areas, noise complaints by local residents began to emerge in the media [3, 4, for example] and in scientific journals [5, 6, for example]. And yet, the vast majority of noise measurements performed in and around homes near wind power plants (WPP) showed values well within the established guidelines [7, 8, for example]. This apparently paradoxical situation has even prompted some authors to assume a psychosomatic origin for resident noise complaints [9], or to associate these health complaints with a lack of monetary gain from the WPP [10]. In direct contradiction to the notion of a psychosomatic origin for these noise complaints, are the animal studies showing increased physiological stress when living in the wild, close to WPP [11, for example], or under laboratory conditions, simulating occupational environments [12, 13, for example].

1.2 Frequency-weighting systems, spectral segmentation and temporal resolution as applied to acoustical data acquisition

The ability of the human auditory system to capture sound depends on the combination of the amplitude of the pressure wave (usually evaluated in deciBels, dB, referenced to 20 microPascal), and the frequency (Hz). Different frequencies require different levels of sound pressure in order to be heard. Some decades ago, the International Organization for Standardization (ISO) established a frequency-weighting network that simulated the human hearing threshold and that was specifically focused

No.	Premise	Sound pressure levels, dB, in octaval bands of averaged geometric frequencies, Hz				General sound pressure level dB "Lin"
		2	4	8	16	
1.	Different jobs inside industrial premises and production areas: - Different physical intensity jobs - Different intellectual emotional tension jobs	 100 95	 95 90	 90 85	 85 80	 100 95
2.	Populated area	90	85	80	75	90
3.	Living and public premises	75	70	65	60	75

Figure 1.
Permissible levels for infrasonic exposures (at 2, 4, 8 and 16 Hz) for two occupational and two environmental settings. Values are provided in dB Linear (no weighting) and, as expected, are lower for public areas than for occupational environments [reproduced from 2].

on preventing hearing loss—the "A" frequency-weighting system [14]. The use of the A-weighting system yields sound pressure levels in the dBA metric.

ISO has also ratified procedures for evaluating infrasound and lower-frequency components: ISO 7196:1995(E) defines the "G" frequency-weighting system as appropriate for quantifying acoustic energy within the range of 0.25–250 Hz [15]. The use of the G-weighting system yields sound pressure levels in the dBG metric. **Figure 2** compares data to which A- and G-weighting have been applied. It also shows the values when no weighting is imposed.

The environment shown in **Figure 2** is within a rural home in the proximity of a WPP, and where residents have noise complaints (see Section 2 below, Home 2). In this 10-minute data segment, the average noise level was less than 30 dBA, well within compliance levels for most rural areas around the world. The G-weighting system, while over-evaluating the sound pressure levels within the range of 10–25 Hz, yielded an average noise level of around 55 dBG. In Japan, for example, the limit for infrasound generated by IWT is 92 dBG [16]. The unweighted capture, which measures the actual levels present in the environment, yielded an average noise level above 60 dB. The highest peaks in this environment, measured without weighting, occurred at frequencies below 8 Hz, i.e., below the defined threshold of human audibility. Taken alone, it would seem that these numerical values are insufficient to adequately characterize the instigator of these residents' noise complaints.

In addition to showing the problematic usage of different frequency-weighting systems, **Figure 2** emphasizes two other aspects of noise measurements: the segmentation of the acoustical spectrum into bands of 1/3 of an octave, and the temporal resolution of 10-minute averages, as per ISO guidelines [14]. As for the spectral segmentation, a higher resolution is technologically possible, but the results are considered mostly academic, since practically all tabulated values related to permissible noise exposure levels use 1/3-octave segmentation.

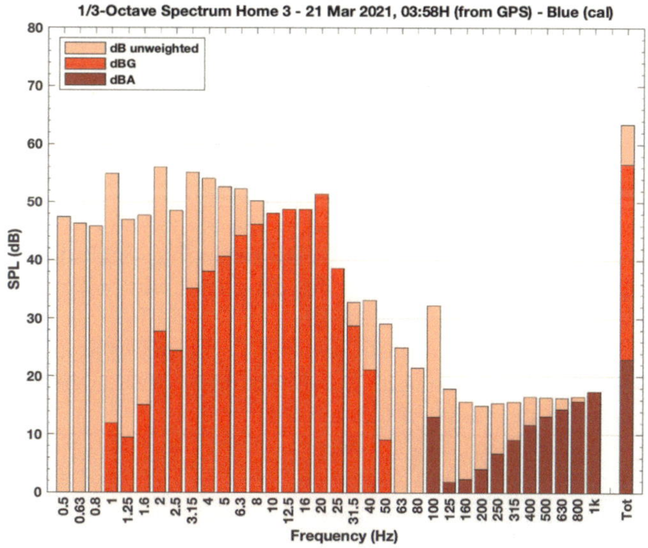

Figure 2.
Comparison of acoustical data acquired with unweighted, G-weighted, and A-weighted systems (10-minute average). Note that between 10 and 25 Hz the G-weighting sound pressure levels are defined to be higher than the unweighted values. (See Section 2 below for detailed methodological capture of this data in Home 2).

1.3 Goal of this report: Going beyond ISO recommendations

Could it be that the spectral segmentation into the 1/3 of an octave and the 10-minute average temporal resolution are too coarse and rudimentary to identify biologically-relevant acoustical phenomena, such as those emanating from WPP?

This report documents the acoustical environments captured in homes located near WPP, using a spectral resolution of 1/36 of an octave, and a temporal resolution of 1-second. Sound pressure levels were analyzed in dB (unweighted).

2. Background and methodology of data collection

Data reported herein were collected in Portugal in Jul-Aug 2020 (Home 1) and in Scotland in Feb-Mar 2021 (Homes 2 and 3), at the invitation of the separate homeowners—usually due to the onset of a pattern of debilitating symptoms which, they claim, only began after WPP became operational in their residential areas [17]. A two-channel sound recording device was placed in and around each home with continuous data acquisition over several days. During the sound recordings, residents were asked to keep a date- and time-logged diary detailing the onset or absence of symptoms, such as sleep disruption. This onset or absence of symptoms could then be compared with changes in the sound recordings that might suggest a causal connection.

2.1 High resolution sound recording

The recording equipment was a SAM Scribe Full Spectrum (FS) system (Soundscape Analytics, Palmerston North, New Zealand), Model Mk1 in Portugal and Mk2 in Scotland [18]. It is a two-channel device with sampling rates up to 44.1 kHz, that is designed to capture recordings of sonic environments with high precision, especially in the infrasonic and low-frequency bands. Data streams are delivered via USB to a Windows notebook computer and stored as uncompressed wav files to a hard disk. GPS information is stored in the files as metadata, which also include a digital signature. Each wav file corresponds to a 10-minute (600-seconds) recording of the sonic environment. The system can accurately record from 0.1–1000 Hz, as per the manufacturer frequency response of the two electret condenser microphones [19].

All measurements reported here cover the range from 0.5–1000 Hz and were captured with a sampling rate of 11.025 kHz. All recordings included a standard reference calibration tone at the start and end, produced with a Type I calibrator (part of the SAM Scribe system) at 1000 Hz/94 dB.

Calibration of the SAM Scribe system rests on 1) the manufacturer's frequency-response curve for the microphone and 2) calibration against a certified Larsen-Davis 831 sound level meter in the range of 6.3–1000 Hz.

2.2 Homes where recordings were captured

2.2.1 Home 1: Portugal (the E. family)

Period of continuous recording: 18 Jul 2020 (00:00)—09 Aug 2020 (10:00). *Microphone location:* At the foot of the bed in master bedroom (ground floor), tripod-mounted 1.5 m above the floor.

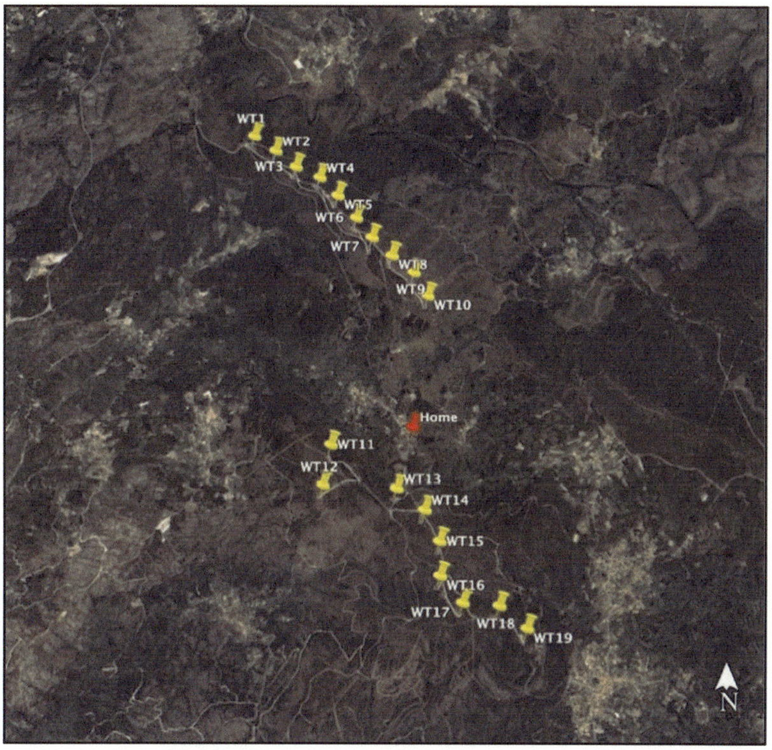

Figure 3.
Relative positions of Home 1 and the 19 industrial wind turbines (labeled WT in the figure) that constitute this wind power plant. (Image adapted from Google Earth).

Figure 3 shows the relative position of Home 1 and the WPP (19 Senvion MM92 turbines of 2 MW each, with blade length 45.2 m). The closest IWTs to the home are numbers 11, 12, 13 and 14, at 843 m, 1085 m, 648 m, and 844 m, respectively. IWT1 and IWT19 are the furthest away, at a distance of 3422 m and 2282 m, respectively.

The E. family—Mr. E. (age 63) and Mrs. E. (age 64)—have lived amid these 19 IWT since 2016. Their health deterioration has been documented by neurological medical reports.

2.2.2 Home 2: Scotland (the P. family)

Period of continuous recording: 24 Feb 2021 (17:30)—07 Mar 2021 (00:00).

Microphone location: Beside the head of the bed in an upstairs bedroom with a dormer, tripod-mounted 1.5 m above the floor.

Mrs. P documented some of her symptoms from Jul 2019 to Mar 2020. **Table 1** shows a 6-month sample (Jul–Dec 2019).

2.2.3 Home 3 – Scotland (The J. Family)

Period of continuous recording: 20 Mar 2021 (16:20)—27 Mar 2021 (18:40).

Microphone location: Middle of attic bedroom, tripod-mounted 1.5 m above the floor.

Symptom	Dates on which symptom was reported
Nausea	6 Jul, 3 Aug, 18 Aug, 12 Oct, 20 Oct, 4 Nov, 6–7 Nov, 10 Nov
Dizziness	7 Jul, 3 Aug, 13–14 Sep, 20–21 Sep, 26 Sep, 28 Sep, 24 Nov, 14–16
Pain in ears	5–9 Jul, 15 Jul, 18 Jul, 22 Jul, 26 Jul, 31 ul, 1 Aug, 3 Aug, 9–12 Aug, 21 Aug, 23 Aug, 13–14 Sep, 2 Oct, 4–5 Oct, 10 Oct, 17 Nov, 22 Dec, 27 Dec, 30 Dec
Sleep disturbance	2 Jul, 4 Jul, 14 Jul, 18 Jul, 22 Jul, 24 Jul, 13 Aug, 25 Aug, 13 Sep, 20 Sep, 12 Oct, 15 Oct, 3–5 Nov, 17 Nov, 23 Nov

Table 1.
Six-month sample of some of the symptoms documented by Mrs. P.

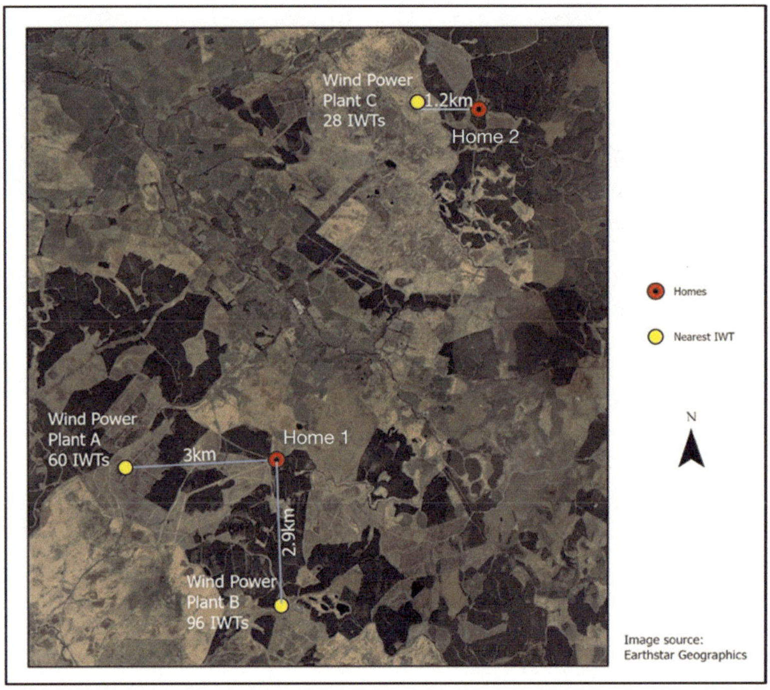

Figure 4.
Relative positions of Home 2 and Home 3 and the closest industrial wind turbines of wind power plants A, B and C.

Figure 4 shows the relative positions between Homes 2 and 3, and the three WPP located in the vicinity.

WPP A has 60 Gamesa G80/2000 turbines of 2 MW each, with blade diameters of 80 m. It is located 4.6 km to the west of Home 2 and 14.5 km to the southwest of Home 3. It has been operational since 2011.

WPP B has 96 Gamesa G114/2500 turbines of 2.5 MW each, with blade diameters of 114 m. It is located approximately 2.9 km to the south of Home 2 and 13.1 km to the south of Home 3. It has been operational since 2007.

WPP C has 28 Gamesa 87/2000 of 2 MW each, with blade diameters of 87 m. It is located approximately 9.5 km to the north of Home 2 and 2.1 km to the southwest of Home 3. It has been operational since 2011.

2.3 Wind data

Information on wind speed and direction was retrieved for the entire period during which recordings were made.

In Portugal, data was obtained from the Portuguese Institute of Sea and Atmosphere (IPMA [20]). Data points were requested in 10-minute increments, from three distinct meteorological stations: at 58 km (altitude above sea level: 995 m), 12.5 km (altitude above sea level: 642 m) and 52.7 km (altitude above sea level: 558 m) away from the E. family home (altitude above sea level: 850 m). In Scotland, weather data was obtained from the British National Weather Institute via the Open Weather service [21] in one-hour intervals. The location for which weather data was obtained was 3.5 km away from Home 2 and 7.8 km from Home 3. Wind data was time-matched to the GPS time-stamped acoustical recordings.

3. Results

3.1 Home 1: Diary

The E. family kept a diary from 13 July through 31 July, 2020.

On 29 July at 04:00, the family's sleep had been disrupted for several hours and Mr. E. felt so unwell that he was compelled to take medication (benzodiazepine) (*'Severe' episode*). By comparison, on the morning of 22 July, Mr. and Mrs. E. slept uninterruptedly until 07:00 (*'Peaceful' episode*).

Priority was therefore given to the analysis of the period between 03:00 and 06:00 (eighteen 10-minute recordings) on both these days, the choice of identical diurnal periods helping to alleviate any extraneous differences between the two mornings.

3.2 Home 1: At 03:20 on the morning of the 'severe' episode (29 Jul, 2020)

Figure 5 shows the results of the sound data acquired between 03:20 and 03:30, on the morning of 29 July, when the E. family's sleep was disrupted and Mr. E. felt the need to self-medicate.

Figure 5A shows a sonogram reflecting the acoustic environment inside the bedroom over a 10-minute period (600 seconds), with 1/36th-octave-band resolution (vertical axis) and 1-second temporal resolution (horizontal axis). The sound pressure level at each frequency and at each second in time, is indicated in the color-coded scale on the right (measured in dB). The yellow color of the straight, horizontal lines visible across the image at 1.5 Hz, 2.3 Hz, 3.0 Hz, and 3.8 Hz reflect the large amount of acoustic energy (50–60 dB) present at these frequencies. Additionally, the lack of discontinuities in these lines indicate that the phenomena were continuously present during the entire 600 seconds.

Figure 5B shows the same numerical data as in **Figure 5A**, but as a frequency spectrum. A series of peaks is readily identifiable, occurring at the same frequencies as the continuous, horizontal lines seen in the sonogram (**Figure 5A**). The mathematical relationship between the frequencies of each peak (red dots) constitutes a harmonic series with a fundamental frequency of 0.76 Hz (0.8 Hz in the figure).

In all 18 recordings (from 03:00 to 06:00, 29 Jul), the sonograms presented similar, continuous horizontal lines and, in all corresponding spectrograms, the same harmonic series (fundamental at 0.76 Hz) was visible. The blade-pass frequency of the IWT

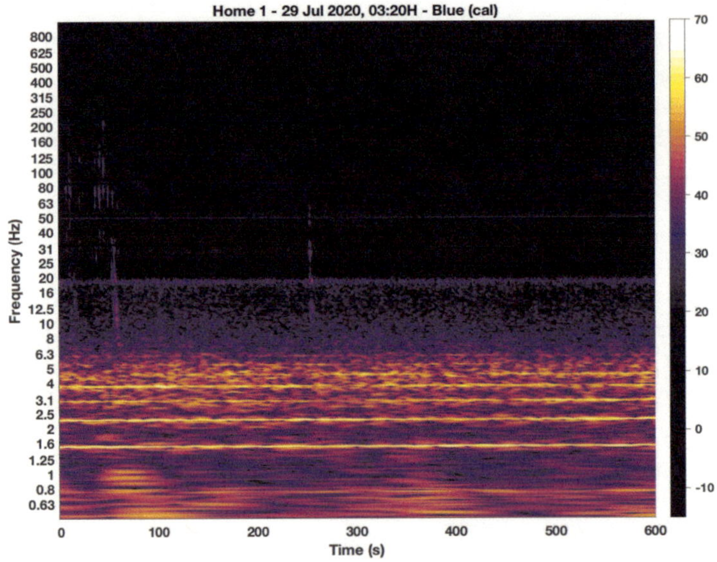

(A)

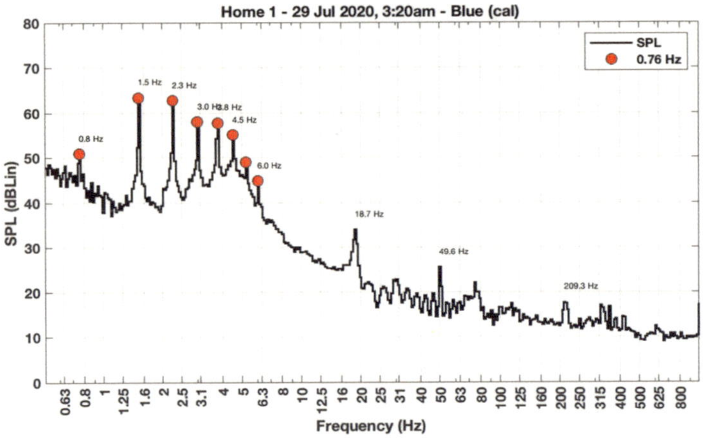

(B)

Figure 5.
(A) Sonogram *showing the sonic environment inside the master bedroom of home 1 (on 29th Jul when sleep was disrupted and medication was required) over a 10-minute period (600 seconds), with 1/36-octave band resolution ('frequency' on vertical axis) and 1-second temporal resolution ('time' on horizontal axis). The color-coded scale on the right measures sound pressure level in (unweighted) dB. Continuous (over the entire 600-second interval), horizontal lines cross the image at 1.5 Hz, 2.3 Hz, 3.0 Hz, and 3.8 Hz with a pressure level of 50–60 dB.*
(B) Spectrogram *in the form of a frequency distribution, constructed with the same numerical data as in* ***Figure 5A****. A harmonic series is identified when the frequencies of each peak (red dots) are multiples of the fundamental frequency of 0.76 Hz (0.8 Hz in the figure).*

installed around the home of family E. is 0.75 Hz. The harmonic series identified in Home 1 is the acoustic signature that emanates from these machines, and that reflects the airborne propagation of a pulsed, pressure wave generated by rotating IWT blades. This IWT acoustic signature occurs below the threshold of human audibility.

3.3 Home 1: At 04:00 on the morning of the 'peaceful' episode (22 Jul, 2020)

In **Figure 6**, the sonic environment in the master bedroom of family E. is shown, as captured between 04:00 and 04:10, on the morning of 22 July, when the E. family slept peacefully. The lack of continuous, horizontal lines throughout the sonogram (**Figure 6A**) is notable, as is the absence of regular peaks in the corresponding

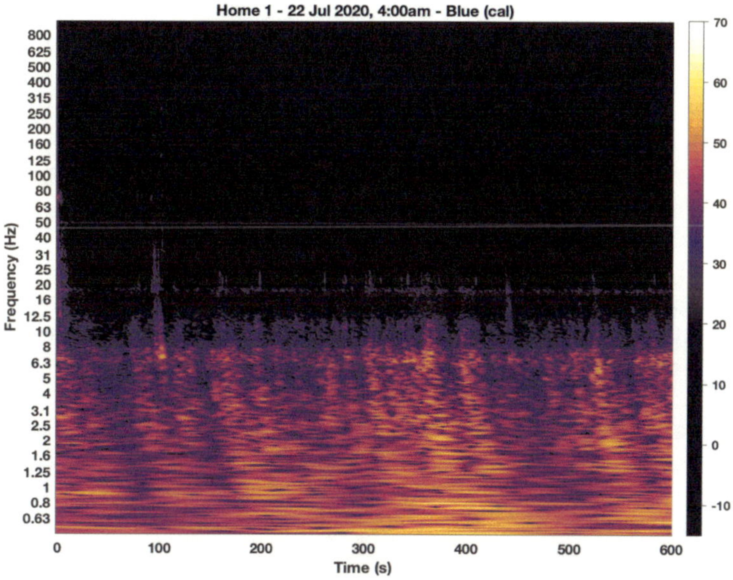

(A)

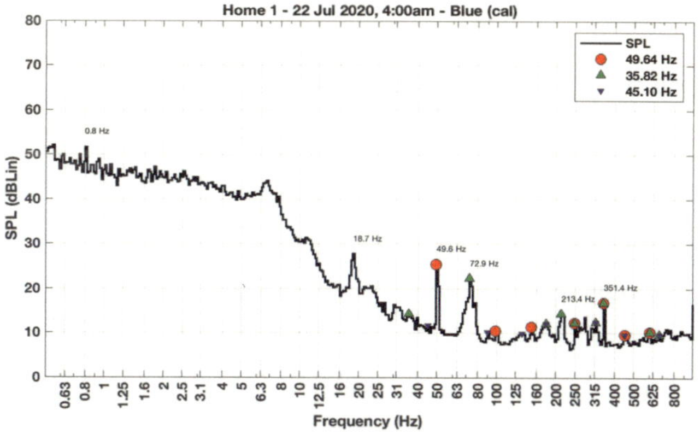

(B)

Figure 6.
(A) Sonogram *showing the sonic environment inside the master bedroom of Home 1 (on 22nd Jul when no sleep disruption occurred) over a period of 600 seconds—With 1/36-octave band resolution, 1-second temporal resolution—and pressure levels in dB as indicated by the color-coded scale. The triangular, pink shapes that span various frequencies are due to blowing wind, and do not exceed 50 dB. Continuous, horizontal lines as observed in **Figure 5A** are absent. (B)* Spectrogram *without any regular, large peaks of acoustic energy in the infrasonic range. Harmonic series, as related to IWT acoustic signatures, are absent.*

spectrogram (**Figure 6B**). The triangular, pink shapes that span various frequencies in the sonogram are due to blowing wind, and do not exceed 50 dB. In all 18 recordings (from 03:00 to 06:00, 22 Jul), no IWT acoustic signature was identified.

3.4 Homes 2 and 3

Regrettably, the residents of these Homes were not sufficiently assiduous with their diary entries so that health-related information could be compared with simultaneous recordings.

Homes 2 and 3 have three different models of IWT among the 3 WPP located in their vicinity, as opposed to Home 1 that only had one type. For asynchronous (constant with varying wind speeds) IWTs, each model will have its own blade-pass frequency and, therefore, their acoustic signatures will be different.

Figure 7 shows the sonogram and spectrogram of the sonic environment captured in the attic bedroom in Home 3. The very clean and continuous horizontal lines that extend throughout the 600-second recording (**Figure 7A**) reflect the existence of a prominent IWT acoustic signature. This is confirmed by the sequence of peaks that constitute the harmonic series, as can be clearly identified in the corresponding spectrogram (**Figure 7B**). The two harmonic series (i.e., IWT acoustic signatures) identified in Home 3 are also present in Home 2, as can be seen in the spectrogram in **Figure 8**.

Figures 7B and **8** show very similar examples of dominant IWT acoustic signatures. The harmonic analysis highlights a harmonic series with a fundamental frequency of 1 Hz (0.99 Hz in the figures) and at least the first 19 harmonics. The Gamesa 80 and 87 IWT models have a blade-pass frequency of 1 Hz. A second harmonic series is identified with a fundamental frequency of 0.67 Hz. The blade-pass frequency for the Gamesa 114 model is 0.67 Hz. A separate harmonic series begins at 20 Hz from an unknown source, possibly the IWT gearboxes.

4. Discussion

4.1 Sleep disruption and the prominence of harmonic peaks

The harmonic series observed in all 18 samples of the 'severe' episode, and that were absent in all 18 samples of the 'peaceful' episode, is recognized as the IWT acoustic signature with a blade-pass frequency of 0.75 Hz. The acoustic signature generated by an IWT is a train of pressure pulses, with a period equal to the reciprocal of the blade-pass frequency of the IWT. It presents as a harmonic series of peaks in the infrasonic region of a spectrogram, visible in **Figures 5B, 7B** and **8**, while absent from **Figure 6B**. In the sonograms, the IWT acoustic signature is present as continuous horizontal lines, as seen in **Figures 5A**, and **7A**, while absent from **Figure 6A**.

This new, high-resolution methodology for assessing infrasonic environments is analogous to transitioning from a magnifying glass to a microscope. Previously undetected acoustic events are now identifiable and, even, quantifiable (see Sections 5.2 and 5.3 below). What was undetectable—and thus assumed to be non-existent, presumably justifying a psychosomatic origin for resident complaints—using the classical noise assessment methodologies (1/3-octave band segmentation in 10-minute averages and with sound pressure levels measured in dBA or dBG), became visible with high-resolution observations.

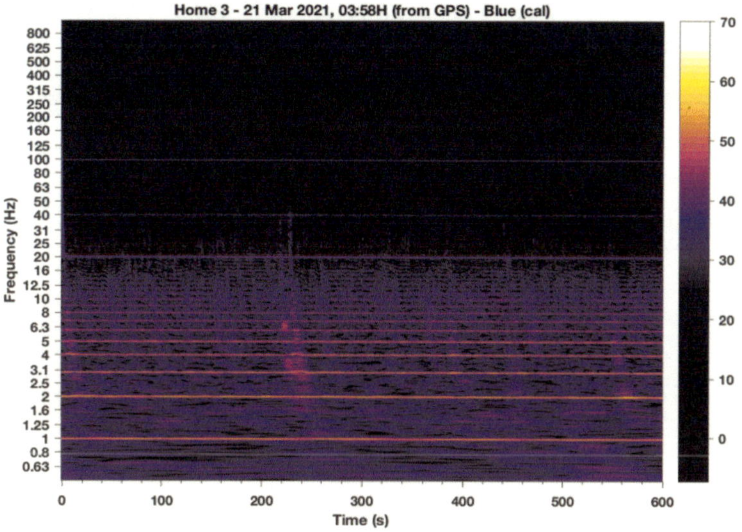

(A)

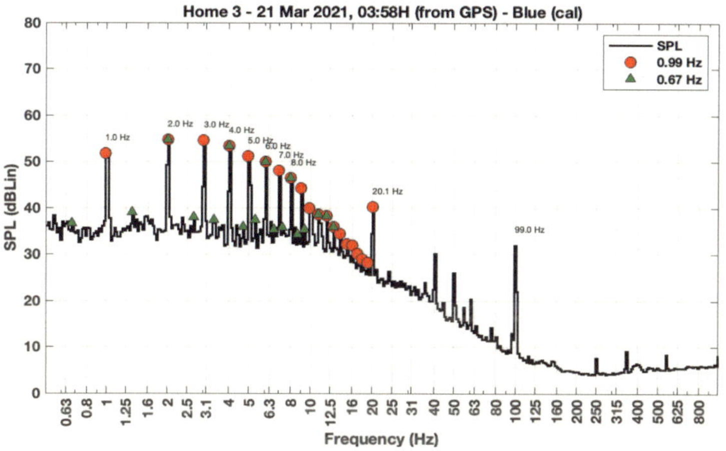

(B)

Figure 7.
(A) Sonogram *showing the sonic environment inside the attic bedroom of Home 3 over a period of 600 seconds, with 1/36-octave band resolution, 1-second temporal resolution, and pressure levels in dB, as indicated by the color-coded scale. Continuous, horizontal lines are readily observable at frequencies below the threshold of audibility, and that reflect the existence of IWT acoustic signatures. (B)* Spectrogram *showing the two most prominent harmonic series, with fundamental frequencies at 0.67 Hz and 0.99 Hz, reflecting IWT acoustic signatures from different IWT models, with different blade-pass frequencies.*

Despite being at frequencies and sound pressure levels that are classically considered as 'below the human hearing threshold,' a very clear correlation has been shown between the existence of these peaks in the frequency spectra and disruption of the normal biological function—sleep disruption followed by the need for self-medication with benzodiazepines. Nevertheless, while the correlation is very clear, the confidence of the correlation is reduced by the relatively small timeframe. Improved confidence

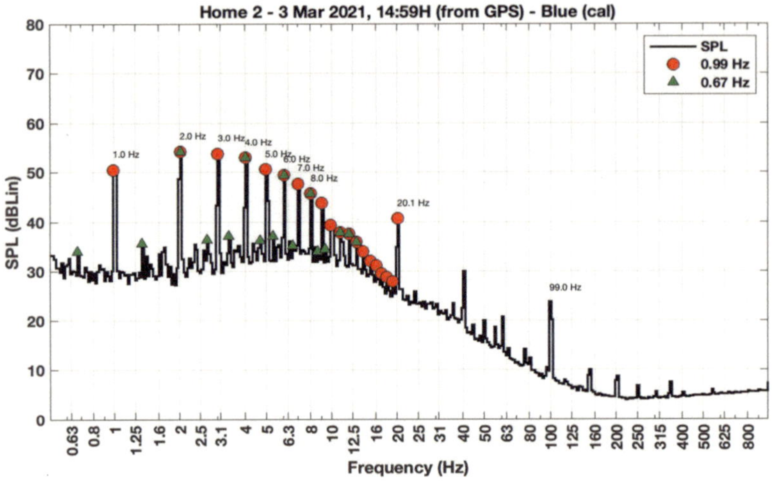

Figure 8.
Spectrogram *showing the sonic environment inside the upstairs bedroom of Home 2, over a period of 600 seconds, with 1/36-octave band resolution and 1-second temporal resolution. Two of the most prominent harmonic series are readily identifiable, with fundamental frequencies at 0.67 Hz and 0.99 Hz, reflecting IWT acoustic signatures from different IWT models, with different blade-pass frequencies.*

can only come from more work to extend the use of this measure to many other cases (an ongoing endeavor by these authors).

The question as to how these infrasonic acoustic events can cause the biological disruption is still unclear. Studies by German scientists, however, using functional magnetic resonance imaging—while exposing subjects to infrasound—may have uncovered a significant clue: in addition to activating the classically identified auditory pathways, infrasonic stimuli also activate regions of the brain that are considered responsible for emotional and autonomic responses [22].

4.2 Prominence of the harmonic peaks—A new metric?

The prominence of these harmonic peaks above the background noise appears to be highly relevant for health-related issues. **Figure 9** depicts a harmonic series as identified in an IWT acoustic signature, an airborne train of pulses occurring within the 0.5–5 Hz window. Note that the persistent or continuous existence of this type of harmonic series ties this acoustic event to human-made sources because the manifestations of such harmonic series from natural sources are exceedingly rare. There is no established methodology to quantify the prominence of these peaks.

A new metric is herein suggested; one that may more accurately provide a measure of the "dose" of this pulsed agent of disease. We have called this measure the *Harmonic Prominence*, H_p, defined as the largest prominence of any harmonic frequency of any harmonic series, within the 0.5–5-hertz frequency window. In **Figure 9**, H_p = 17 dB, at 1.5 Hz. In the specific case of IWT, only harmonic series with a fundamental frequency equal to the IWT blade-pass frequency are considered. In the specific case of the data acquisition methodology detailed above, the highest prominence of the harmonic series is determined in temporal segments of 600-seconds.

There are a variety of mathematical definitions, methodologies and software packages associated with quantifying peak prominence above background, for almost any and all types of wave phenomena. These authors have adhered to the formal

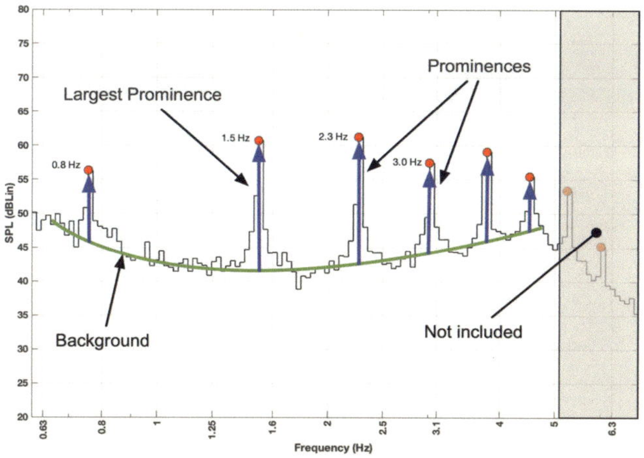

Figure 9.
Determination of prominence levels based on 1/36-octave frequency bands. The largest prominence, H_p (see text), in this series is approximately 17 dB over background. (Numerical data for this figure were obtained in Home 1, during the 'severe' episode).

definition of prominence in a frequency spectrogram as established by MATLAB which has a robust definition of prominence in terms of the peak height and the local background level [23].

The H_p parameter does not measure the total energy of the pulses in the pulse train that emanates from IWT. This energy is spread out over all the harmonic components of the pulses—the peaks in the spectrogram—whereas the measure only looks at the peak with the largest prominence. Therefore, H_p cannot be considered as an energy measure.

Another approach would be to look in the time domain, rather than in the frequency domain. Here a measure such as the crest factor could be used to gain a measure of the 'peakiness' of the pulses, using their total energy. These additional avenues of research are undergoing further scrutiny by these authors and their colleagues [17].

4.3 Day-time plots—Evaluation of long-term infrasound exposures

The H_p parameter can provide health scientists with a rudimentary indicator of the largest prominence above background that exists within a 10-minute measurement. When continuous measurements are maintained over several days (or weeks), a clearer picture regarding the long-term variation of exposure to these trains of pulses is revealed.

Figure 10 shows a Day-Time plot for the data collected in Home 1, 18 Jul-09 Aug, 2020. Here H_p is plotted as a surface with the date as the abscissa and the time of day as the ordinate. For each 24-hour period, there are 144 ten-minute samples. The values of H_p were determined for each 10-minute sample, and then binned (scale: <5 dB, 5–10 dB, 10–15 dB, 15–20 dB, 20–25 dB and > 25 dB), as reflected by the color-coded scale in **Figure 10**.

Similar day-time plots were constructed for Homes 2 and 3, as shown in **Figures 11** and **12**, respectively. While these types of plots are informative as to the time and

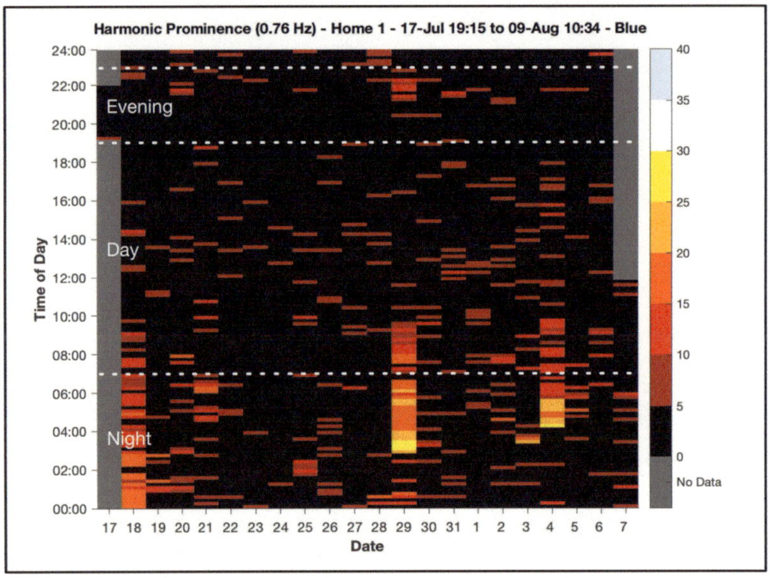

Figure 10.
Day-time plot for Home 1. *The 'severe' episode took place on 29th Jul, while the 'peaceful' episode took place on 22 Jul. The nights of 18 Jul and 4 Aug also show the presence of prominences. The 'peaceful' morning (22 Jul) has only one 10-minute sample with a significant H_p level. The following two mornings also appear to have no significant H_p samples but were not noted in the residents' diary as either peaceful or disturbed. The 'severe' morning (29 Jul), from 3 am until about 9 am shows up in stark contrast to the other mornings, indicating not only that the H_p levels were high but also that they were the highest in the entire length of the recording. The night of 4 Aug also shows an interval of 10-minute samples with severe H_p levels. Since the residents' diary stops on 31 Jul their experience on this day was not recorded. Finally, the night of 18 Jul shows elevated H_p levels from midnight onwards, although these did not reach the same levels as for 29 Jul or 4 Aug. The Es' diary entry for 18 Jul at 04:00 indicated that the "noise was unbearable" and "sounded like a derailing train."*

duration that people are exposed to higher or lower levels of H_p, it is still important to view the sonograms to get a true understanding of the nature of the sonic environment at that point in time. For example, it is not possible from this graph alone to determine if the lower H_p levels seen in Home 2 on the morning of the 27th (**Figure 11**) are caused by the presence of a higher background noise level or whether the levels of H_p were actually diminished.

Note that not all the 10-minute intervals where the H_p is shown as 0 (black) are, in fact, 0. Impulsive sound—caused by such events as people walking over a floor or a door closing—can contaminate an entire 10-minute recording since the impulse is spread over longer and longer time intervals as the frequency of the 1/36-octave bands decrease.

To use the H_p measure as part of a dose–response metric, the simplest method would be to integrate it over time, i.e., multiply each value by 10 minutes and sum for a metric in decibel-minutes. Long-term exposure might be measured in decibel-years. Future research might even develop infrasound dosimeters for workers, similar to those used for radiation exposures.

Comparing the infrasonic environment in Home 1 with those encountered in Homes 2 and 3, a major difference becomes obvious: in the latter two homes, periods of respite (black areas in the day-time plots) are almost non-existent. Periods of respite are understood as biological recovery times, during which the agent of disease is not present and physiological cellular repair can be undertaken unimpeded by the acoustic aggressor. In Home 1 there is the possibility of comparison between the

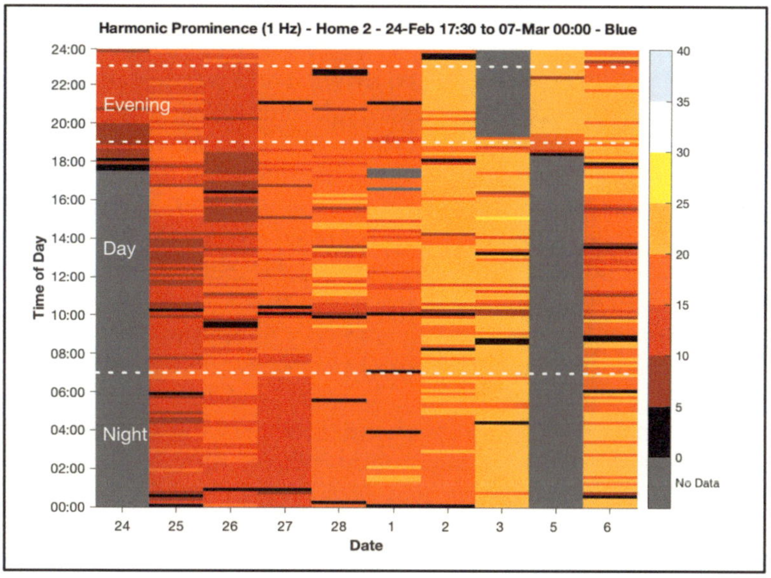

Figure 11.
Day-time plot for Home 2. *A visual inspection shows that the H_p was most dominant from the 2nd through the morning of the 6th reaching its highest value at around 3 pm on the 3rd, with H_p between 25 and 30 dB above background.*

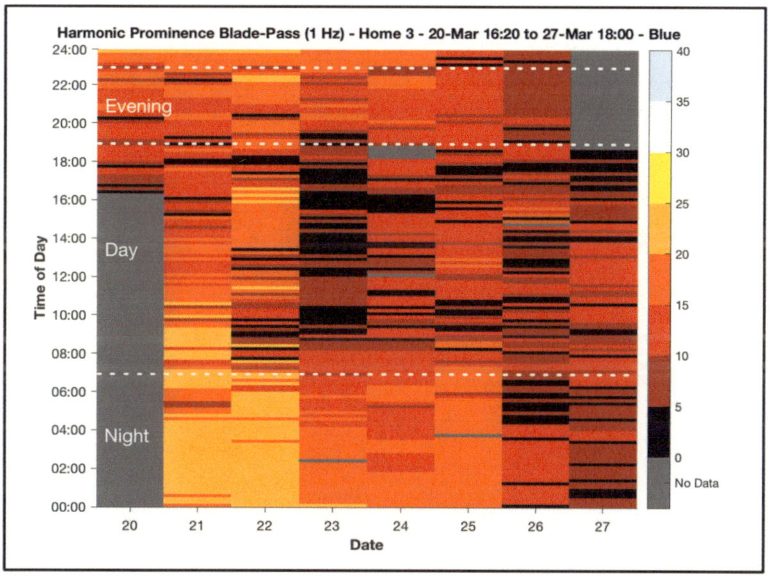

Figure 12.
Day-time plot for Home 3. *The most dominant episodes, i.e., highest level of H_p, were at night. The mornings of the 21st and 22nd registered the strongest H_p (20–25 dB), while the morning of the 27th presented with the weakest.*

periods of time when the IWT acoustic signature is present and when it is absent. Clearly, this is a much more difficult proposition in Homes 2 and 3, where the H_p level indicates that IWT acoustic signatures are almost always present, to a greater or lesser extent (color-coded scale).

4.4 Harmonic prominences wind roses

Airborne sound propagation is affected by wind and weather conditions. In addition to the obvious fact that wind 'carries sound,' thus reducing attenuation downwind, other atmospheric properties can greatly alter both the propagation and attenuation of sound. For instance, increasing humidity will improve propagation, while atmospheric inversion layers can create 'dead zones' where sounds will not be heard despite proximity to the source. Beyond these effects, the propagation and attenuation of infrasound differs in some important respects from sound at higher frequencies. While higher-frequency sound diminishes by 6 dB per doubling of distance—the inverse-square law—infrasound only diminishes by 3 dB. Infrasound is also more prone to refraction around large objects such as hills and to being funneled down valleys.

A data fusion of meteorological data (wind direction) and acoustic data (H_p) can provide insight into these weather- and terrain-induced differences that can significantly influence H_p levels. A *harmonic prominence wind rose*, which takes its inspiration from the common wind rose, is the nomenclature given to this data fusion. An example can be seen in **Figure 13**, reflecting data obtained in Home 1.

The H_p wind rose is a stacked, frequency histogram plotted in polar coordinates. It shows the number of 10-minute samples with an H_p in each dB-level bin in the direction of the then-prevailing wind. Each bin is identified by a color and the number of samples is indicated by the length of each segment in the radial direction. This provides important information if, for instance, the strongest levels of H_p align with a given wind direction.

The closest national meteorological stations must be used to provide wind data if a certified weather station is not available at the site of the sound recordings. This may be problematic since many weather services do not provide data at the closest weather

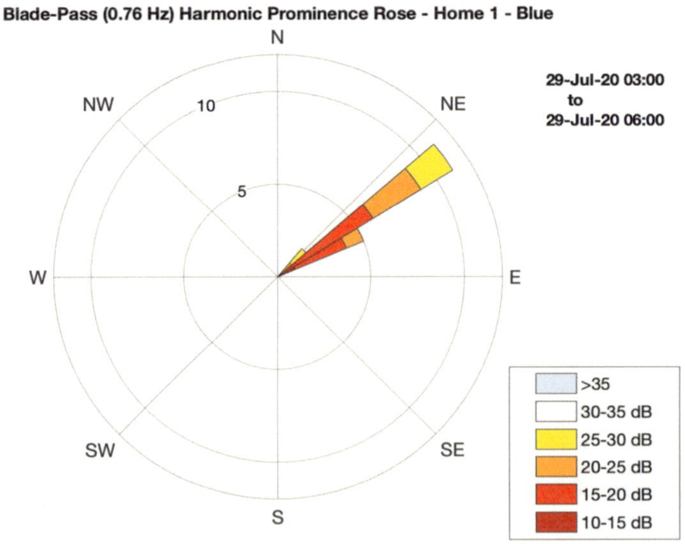

Figure 13.
Harmonic prominence wind rose for Home 1. *Data refers to the 18 samples examined during the 'severe' episode. The highest Hp levels (yellow) were registered when wind was from the north-eastern quadrant. (Wind data from weather station located 12.5 km from Home 1).*

station but rather synthesized weather information, using their proprietary weather models. The wind direction provided is, therefore, not necessarily the same as at the recording site. Moreover, the wind direction at the hub-height of the IWTs may not be the same as at the height of the weather station or the home. The wind direction cannot, therefore, be said to indicate the direction of the source of the IWT acoustic signature in relation to the home.

While the wind direction can provide some understanding, the windspeed may also have a tale to tell. This leads to H_p wind roses plotted for data within wind-speed ranges. Examples are shown in **Figures 14** and **15** for three ranges. The left graph is a H_p wind rose for all 10-minute periods in the recording interval when the wind was between 0 and 10 km/h, the middle graph is for wind speeds of 10–30 km/h and the right graph is for wind speeds of 30–60 km/h.

The fact that the multi-day recording in Home 3 does not include any wind from the eastern half of the compass emphasizes the fact that a reasonable sampling of wind conditions will involve recordings from throughout the year to cover all seasons.

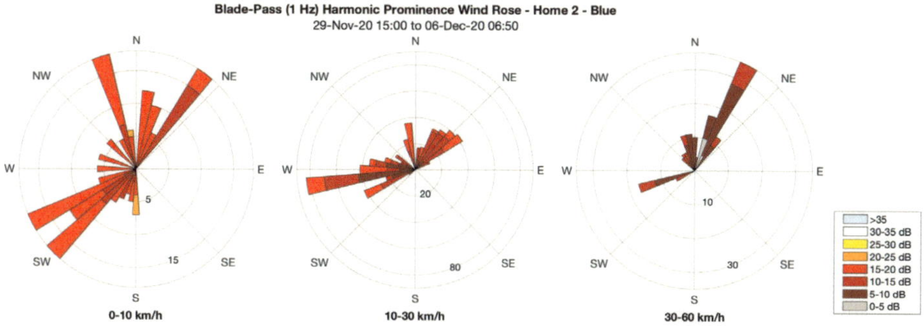

Figure 14.
Harmonic prominence wind roses for three wind-speed ranges for Home 2. *The strongest H_p is at the lowest windspeed, and this is most consistently dominant when the wind is between southwest and north-northeast; i.e., the sectors of the wind rose in these directions are almost entirely made up of 10-minute intervals where the H_p was between 15 and 20 dB (red). By comparison, where the wind was from the northeast, only about 15% of the sound files have this level. At the highest windspeeds, no instances of 15–20 dB H_p can be seen.*

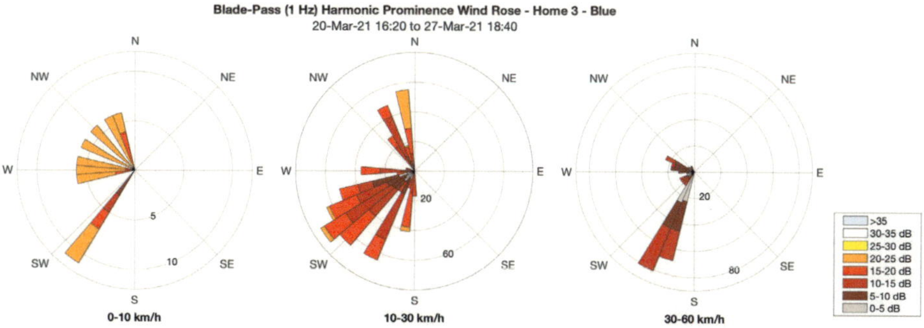

Figure 15.
Harmonic prominence wind roses for three wind-speed ranges for Home 3. *The strongest H_p is at the lowest windspeed and this is most consistently dominant when the wind is from the west through to northwest. That is, the sectors of the wind rose in these directions are almost entirely made up of 10-minute intervals where the H_p was between 20 and 25 dB (orange). By comparison, where the wind was from the southwest, only about 1/3 of the sound files have this level. At the highest windspeeds, no instances of even 15–20 dB H_p can be seen.*

The inverse dependence of H_p on wind speed is because the wind noise increases with wind speed, thus increasing the levels of background noise—including in the infrasound region. The sound pressure levels of the IWT pulses, however, remain constant. Thus, H_p will have lower values (see **Figure 9**). The unanswered question is whether the human brain processes the information of the IWT acoustic signatures when these appear obscured by, or embedded in, the increased background noise, as measured by a machine.

4.5 The position of other authors

In this type of scientific endeavor, it is normally expected that the work of other authors also be presented to form a context and allow a comparative analysis of results obtained and/or of the methodologies used. Regrettably, most, if not all, papers on infrasound are conducted with a ⅓–octave resolution, which immediately precludes any data comparison with that presented here. Due to a variety of conditioning factors that have been in place for decades, sound level meters readily available on the market do not possess the technical capabilities for this type of data acquisition and subsequent analyses. Simultaneously, many of the health-related aspects that are studied within the context of IWT are restricted to measures of "annoyance" (a non-clinical and highly subjective parameter) or to the audibility of the sound, neither of which are very relevant to the results presented here.

In 2018, the World Health Organization (WHO) published a document titled: *Environmental Noise Guidelines for the European Region* [24]. The word "infrasound" has one single entry, on page 85, under the section heading *Wind turbine noise*:

> *"Wind turbines can generate infrasound or lower frequencies of sound than traffic sources. However, few studies relating exposure to such noise from wind turbines to health effects are available. It is also unknown whether lower frequencies of sound generated outdoors are audible indoors, particularly when windows are closed."*

These and other statements reflect a profound misunderstanding of the importance of the time-profile of an exposure to sound as it relates to biological responses (e.g., traffic does not produce harmonic peaks with a one-second pulse rate). However, in defense of this position taken by the WHO, it must be acknowledged that the methodologies it uses for assessing sound necessarily preclude the observation and identification of harmonic series associated with IWT. The suggestion that the audibility of infrasound levels (in itself, an oxymoron by classical definitions) can be mitigated by closed windows clearly indicates a profound lack of knowledge on the physical attributes of propagating airborne pressure waves within the infrasonic range [25–27].

5. Conclusions

This chapter provides a different approach to the measurement and analysis of infrasound in and around homes located in the proximity of wind power plants. Examples show how using higher temporal- and spectral-resolutions (1 second and 1/36 of an octave), and without any frequency weighting, can reveal acoustical features in the infrasonic range that may indicate a causal relationship with self-reported medical symptoms. This possibility is usually considered non-existent since the infrasonic range is generally viewed as inaudible, and thus innocuous, to humans. The suggestion therefore arises that current noise protection procedures are insufficient to protect public and occupational health. The approach used by these authors offers a more solid framework with which to pursue the establishment of dose–response relationships for infrasonic exposures. Future studies are being extended into noisy occupational environments and different environmental settings where wind power is not the acoustic source.

Acknowledgements

We would like to thank the invaluable assistance of Susan Crosthwaite (Citizen's Initiative UK and Independent Noise Working Group – INWG) and Melvin Grosvenor (Grosvenor Consultancy and Independent Noise Working Group – INWG). We would also like to thank all the families who have contributed and continue to contribute to this Citizen's Science Initiative for the Acoustic Characterization of Human Environments. Finally, we would like to express our deep appreciation for the fundamental knowledge and long-time support provided by Dr. Bruce I. Rapley and Soundscape Analytics.

This project has received ethical approved from the NZ Ethics Committee (see www.nzethics.com), application number NZEC19_12.

Conflicts of interest

HHCB developed software for capturing and analyzing the sound files for the SAM system, no financial interest. MAP no conflict. RM no conflict. RS contributed to the development of the SAM system, no financial interest. PD no conflict.

Author details

Huub Bakker[1], Mariana Alves-Pereira[2*], Richard Mann[3], Rachel Summers[1] and Philip Dickinson[1]

1 International Acoustics Research Organization (IARO), Palmerston North, New Zealand

2 Lusófona University, Lisbon, Portugal

3 University of Waterloo, Ontario, Canada

*Address all correspondence to: m.alvespereira@gmail.com

References

[1] Pimorov L, editor. Les Infra-Sons. France: CNRS Publishing; 1974

[2] Stepanov V. Biological Effects of Low Frequency Acoustic Oscillations and Their Hygienic Regulation. Moscow: State Research Center of the Russian Federation; 2000

[3] Kaeding EF. The curse of repowering – A long descent. Die Tageszeitung. 2014. [In German] https://www.taz.de/Archiv-Suche/!5032786&s=hogeveen/

[4] Wetzel D. [Energy Danish Debate - Does the infrasound of wind turbines make you sick?]. Die Welt. 2015. [In German] https://www.welt.de/wirtschaft/energie/article137970641/Macht-der-Infraschall-von-Windkraftanlagen-krank.html

[5] Dumbrille A, McMurtry RY, Krogh CM. Wind turbines and adverse health effects: Applying Bradford Hill's criteria for causation. Environmental Disease. 2021;**6**:65-87

[6] Zajamsek B, Micic G, Hansen K, Catcheside DPN. Wind farm infrasound detectability and its effects on the perception of wind farm noise amplitude modulation. In: Proceedings of the Annual Conference of the Australian Acoustical Society; 10–13 November 2019. Cape Schanck, Victoria, Australia; 2019. pp. 487-494

[7] Maijala P, Turunen A, Kurki I, Vainio L, Pakarinen S, et al. Infrasound Does Not Explain Symptoms Related to Wind Turbines. Helsinki, Finland: Publication Series of the Government's Study and Research Activities; 2020. p. 34

[8] Ratzel U, Bayer O Brachat P, Hoffmann M, Janke K et al. Low frequency noise including infrasound from wind turbines and other sources: Report on results of the measurement project 2013-2015. State Office for the Environment, Measurement and Nature Conservation of the Federal State of Baden-Wuerttemberg, Karlsruhe, Germany. 2020. https://pudi.lubw.de/detailseite/-/publication/13796-Report_on_results_of_the_measurement_project_2013-2015.pdf

[9] Crichton F, Chapman S, Cundy T, Petrie KJ. The link between health complaints and wind turbines: Support for the nocebo expectations hypothesis. Frontiers of Public Health. 2014;**2**: Article 220

[10] Pedersen E, van den Berg F, Bakker R, Bouma J. Response to noise from modern wind farms in The Netherlands. Journal of the Acoustical Society of America. 2009;**126**:634-643. DOI: 10.1121/1.3160293

[11] Agnew RCN, Smith VJ, Fowkes RC. Wind turbines cause chronic stress in badgers (meles meles) in Great Britain. Journal of Wildlife Diseases. 2016;**53**: 459-467

[12] Zou LH, Shi YJ, He H, Jiang SM, Huo FF, et al. Effects of FGF2/FGFR1 pathway on expression of A1 astrocytes after infrasound exposure. Frontiers in Neuroscience. 2019;**13**:429. DOI: 10.3389/fnins.2019.00429

[13] Zhao JH, Wang JH, Luo JY, Guo XY, Wang Y, et al. Effects of infrasound on gastric motility, gastric morphology and expression of nitric oxide synthase in rat. Biomedical and Environmental Sciences. 2018;**31**:399-402. DOI: 10.3967/bes2018.052

[14] ISO1996-2:2007(E). Acoustics. Description, Measurement and

Assessment of Environmental Noise. Part 2: Determination of Environmental Noise Levels. Geneva, Switzerland: ISO; 2007

[15] ISO 7196:1995(E). Acoustics. Frequency-weighting Characteristic for Infrasound Measurements. Geneva, Switzerland: ISO; 1995

[16] Health Protection Agency (UK). Health Effect of Exposure to Ultrasound and Infrasound—Report of the Independent Advisory Group on Non-ionising Radiation. London: Health Protection Agency; 2010

[17] IARO-International Acoustics Research Organization-represents a group of scientists who, collectively, hold over 200 years of scientific experience in the field of infrasound and low frequency noise, and its effects of human health. Since 2016, our researchers have been recording and analysing acoustical data in and near homes located in the vicinity of onshore wind power stations, in the following countries (alphabetical): Australia, Canada, Denmark, England, France, Germany, Ireland, New Zealand, Northern Ireland, Portugal, Scotland, Slovenia, and The Netherlands. Prior to 2016, all IARO scientists were already working either in acoustics alone or in acoustics and health. All research conducted by IARO is part of the Citizen Science Initiative for Acoustic Characterization of Human Environments (CSI-ACHE), the research protocols for which have been approved by the New Zealand Ethics Committee (application number NZEC19_12). www.iaro.org.nz

[18] Bakker HHC, Rapley BI, Summers SR, Alves-Pereira M, Dickinson PJ. An affordable recording instrument for the acoustical characterisation of human environments. In: Proceedings International Conference Biological Effects of Noise (ICBEN). Zurich, Switzerland; 2017

[19] Model No.: EM246 ASSY, Primo Co, Ltd, Tokyo, Japan. Available from: https://www.primo.com.sg/components/

[20] IPMA [Portuguese National Institute for the Sea and Atmosphere]. www.ipma.pt

[21] Open Weather, London, UK, 2022. http://www.openweathermap.org

[22] Weichenberger M, Bauer M, Kuhler R, Hensel J, Forlim CG, et al. Altered cortical and subcorticol connectivity due to infrasound administered near the hearing threshold – Evidence from fMRI. PLoS One. 2017;**12**:e0174420. DOI: 10.1371/journal. pone.0174420

[23] The Mathworks, Natick, USA, v2020b. It should be noted that the Matlab findpeaks function can sometimes return significantly higher prominences than their definition allows. Another method was developed to derive the appropriate prominence level: IARO21-3. White Paper on the Harmonic Prominence Measure. IARO, Palmerston North, New Zealand. Available from: https://iaro.org.nz/wp-content/uploads/2022/01/IARO21-3-White-Paper-on-the-Harmonic-Prominence-Measure-v6.pdf.

[24] World Health Organization. Environmental noise guidelines for the European region. 2018. ISBN 978 92 890 5356 3. Available from: https://www.euro.who.int/en/publications/abstracts/environmental-noise-guidelines-for-the-european-region-2018

[25] Alves-Pereira M, Bakker HHC. Occupational and residential exposures to infrasound and low frequency noise in aerospace professionals: Flawed assumptions, inappropriate quantification of acoustic environments, and the inability to determine dose–response. Scientific Journal of Aerospace Engineering and Mechanics. 2017;**1**(2): 83-98

[26] Alves-Pereira M, Krough C, HHC B, Summers R, Rapley B. Infrasound and low frequency noise guidelines – Antiquated and irrelevant for protecting populations. In: Proceedings of the 26th International Conference on Sound & Vibration. Montreal, Canada; 2019

[27] Alves-Pereira M, Rapley B, Bakker HHC, Summers R. Acoustics and biological structures. In: Abiddine ZE, Ogam E, editors. Acoustics of Materials. London: IntechOpen; 2019